Environmental Science

Series editors: R. Allan · U. Förstner · W. Salomons

Springer

Berlin
Heidelberg
New York
Barcelona
Hong Kong
London
Milan
Paris
Singapore
Tokyo

Manfred Vollmer · Henning Grann (Eds.)

Large-Scale Constructions in Coastal Environments

Conflict Resolution Strategies

**First International Symposium
April 1997, Norderney Island, Germany**

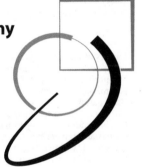

With 51 Figures and 10 Tables

 Springer

Manfred Vollmer
ENCORE (Environmental Conflict Resolution)
Weserstraße 78
D-26382 Wilhelmshaven
Germany

Henning Grann
Statoil Deutschland GmbH
Jannes-Ohling-Straße 40
D-26723 Emden
Germany

ISSN 1431-6250
ISBN 3-540-64647-7 Springer-Verlag Berlin Heidelberg New York

Library of Congress Cataloging-in-Publication Data
Large-scale constructions in coastal environments : conflict resolution strategies : First International Symposium, April 1997, Norderney Island, Germany / Manfred Vollmer, Henning Grann (eds.).
p. cm. -- (Environmental science)
"International Symposium on Large-Scale Constructions in Coastal Environments, held from 21-25 April 1997 on Norderney Island, Germany"--Pref.
Includes index. ISBN 3-540-64647-7 (herdcover : alk. paper)
1. Coastal engineering--Congresses. 2. Coastal zone managament--International cooperation--Congresses. I. Vollmer, Manfred, 1956-. II. Grann, Henning, 1933- . III. International Symposium on Large-Scale Constructions in Coastal Environments (1st : 1997 : Norderney Island, Germany) IV. Series: Environmental science (Berlin, Germany)
TC203.5.L37 1998
333.91'715--dc21

98-34468
CIP

© Springer-Verlag Berlin Heidelberg 1999
Printed in Germany

The use of general descriptive names, registered names, trademarks, etc. in this publication does not imply, even in the absence of a specific statement, that such names are exempt from the relevant protective laws and regulations and therefore free for general use.

Cover Design: Struve & Partner, Heidelberg
Dataconversion: Büro Stasch, Bayreuth

SPIN: 10679754   32/3020 – 5 4 3 2 1 0 – Printed on acid-free paper

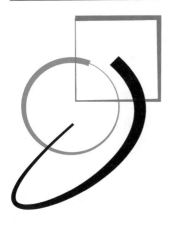

First
International Symposium
on Large-Scale Constructions
in Coastal Environments

held from
21–25 April 1997
on Norderney Island, Germany

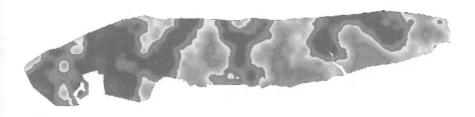

Recommendations
and
Conclusions

Recommendations

Preamble

In most parts of the world coastal ecosystems are vital strongholds of biodiversity and provide important habitats for highly adapted and hence very sensitive plant and animal communities. On the other hand, these land/sea transition zones have been and are increasingly becoming important sites for human activities. This is highlighted by the fact that over 60% of the world's population lives within 50 km of the coast.

The increasing number of large-scale constructions exerts strong pressures on the environment. Our modern technological civilisation has to deal with the problems created by the globalisation of economic activities, recreation demands, an exploding population and urbanisation, the needs of sustainable economic development, and efforts to preserve the environment on a global scale. These pressures on our ecological and social systems inevitably result in conflicts between economy and ecology, and have precipitated discussions about the value of the biophysical environment, and the need for a new ethic.

In order to tackle these problems the Symposium on Large-Scale Constructions in Coastal Environments for the first time brought together spatial planners and specialists in the fields of environmental policy, environmental economy, environmental ethics and environmental science with the objective of identifying and discussing cause/effect relationships, and recommending potential remedies to the numerous conflicts between environmental concerns and economic interests. Solutions require sound knowledge of the biophysical and socio-economic environment as a unified system.

In the course of the conference a number of factors and issues were identified which appear to lie at the root of these conflicts. On this basis, the open participatory process and the interconnection between all disciplines led to a number of important recommendations which primarily focus on the need for developing a comprehensive, integrated environmental management approach.

The recommendations are aimed at all stakeholders, such as heavy industry and other economic businesses, planning organisations and supervising authorities involved in the implementation, execution and management of large-scale constructions, governmental ministries and agencies implicated in such developments, as well as the general public affected by such activities, including non-governmental organisations such as nature conservation bodies and societies.

Although it is beyond the power of the conference delegates to enforce implementation of the proposed recommendations, they call upon the support and assistance of international organisations such as the UNECE and UNEP with the request that they pass them on to governments and other policy makers for their urgent attention.

Recommendations

1. Large-scale constructions have far-reaching impacts on the environment and the communities living in affected areas. Consequently, the planning and implementation of construction schemes should adopt an integrated approach in which the interests of all involved stakeholders are considered and carefully balanced.
2. Every business, organisation and authority involved in the development and implementation of large-scale construction plans should aim for sustainable solutions. Environmental concerns, and protected areas in particular, should be respected by the sharing of responsibilities and by taking voluntary initiatives beyond compliance, in keeping with good corporate citizenship. In their dealings with each other stakeholders should observe basic ethical principles.
3. All stakeholders should be involved in the early planning stages and are encouraged to seek consensus on acceptable impacts and appropriate compensation measures. The public should pro-actively be involved in this process, and independent mediation should be available at all times.
4. Integrated management should be based on rational and clear political decisions entailing well-defined objectives in order to ensure the development of sound management policies and the proper involvement of stakeholders. Furthermore, sound integrated management must take full account of environmental costs in order to provide a firm basis for the decision-making process in the course of project execution.
5. Environmental policy and management procedures should be improved by extending the application of the precautionary principle to physical impacts on habitats, by elaborating implementation schemes for the precautionary principle, taking regional differences into account, and by designing more effective monitoring programs. In addition, proper environmental protection practises should be applied during project execution.
6. Cost-benefit analyses should include all environmental, social, cultural, and transborder-derived effects, with a differential application of the best available technology not entailing excessive costs, and should observe equity principles in the process.
7. Assumptions and predictions of environmental impact assessments (EIAs) should be critically reviewed during construction for the purpose of taking immediate remedial action, as well as after construction for the purpose of improving future EIAs.
8. Since interdisciplinary training and education are indispensable for effective integrated management, appropriate training programs and courses should be developed, and suitable educational facilities should be made available.

Conclusions

Compiled by M. Vollmer, H. Grann and B.W. Flemming

Preface

Like many such conferences, the First International Symposium on Large-Scale Constructions in Coastal Environments took place because there were local and regional concerns about the impacts of economic development on sensitive environments. The focus was on ways to link economic interests and environmental concerns. This requires great efforts, but one also has to accept that there are no unique solutions. In fact, every affected region has to find its own approach to the problem, although guidelines at national and international levels could provide a useful framework.

Considering the subject of the conference it should be recognized that in many cases large-scale projects on the coast evolve through the growth, merger, and linkage of many smaller projects.

Furthermore, measures are often taken to combat short-term detrimental effects, ignoring the fact that sometimes more chronic and long-lasting effects may emerge at a later stage. The cumulative impact of many smaller projects developing into large-scale constructions, therefore, also has to be considered.

The symposium conclusions, which were debated by the conference delegates before being adopted, contain explanations and comments to accompany the more concise recommendations. The conclusions are numbered 1 through 8 and correspond to the eight recommendations.

Conclusions

1. Modern economic development with its large-scale constructions exerts strong pressures on the environment. These pressures on our ecological and social systems very often result in conflicts between economy and ecological/social interests. In order to tackle these problems a pro-active management strategy for environmental conflict resolution on the basis of ethical principles is required.

Consequently, a bold strategy of this type should adopt integrated management plans in which the needs of economic development and the interests of all other involved stakeholders are carefully balanced. An open participatory process acknowledging the interconnections of the various objectives leads to integrated solutions capable of closing the gap between economy on the one hand and ecological and social interests on the other.

Integrated approaches have to be formulated by the involved stakeholders. In this context, project fora are a useful instrument. They should be established at the very beginning of the project planning phase. All interest groups should be invited to participate in such discussion and decision-making fora. The overriding aim should be to reach consensus on the planning design and construction of the economic projects. In this process, alternatives have to be taken into consideration and even zero-options must be possible if strong ethical principles demand this.

The current practise of attaining agreement on project plans highlights two major dilemmas:

1. All involved stakeholders formulate their aims and objectives concerning a new project. A co-ordinating authority or bureau then combines these different interests and formulates a common approach. However, in this process important aspects are commonly overlooked because it is impossible to integrate varied points of view without the participation of the stakeholders themselves. As a result, some interest groups may feel disadvantaged and this, in turn, precipitates new conflicts, which often culminate in confrontation. Integrated approaches can only succeed if the aims and objectives are formulated in an open participatory process. All stakeholders must feel that they made positive contributions to the final outcome. Integrated approaches thus require interdisciplinary initiatives if they are to be successful.
2. Responsibilities are spread throughout local authorities, governments and ministries. Jurisdiction thus constantly needs to be clarified during the different project phases, a process commonly characterised by prolonged discussions about specific areas of responsibility. This leads to delays in decision-making and hence to unsatisfactory results. A concentration of responsibilities in the project fora would greatly improve and accelerate the decision-making process.

2. Our modern technological civilisation has to cope with the need for economic development on a global scale, while at the same time giving due attention to the problems of globalisation of economic activities, tourism, exploding population growth and urbanisation without destroying the environment. According to the Brundtland Commission and to Agenda 21 of the Rio summit, every business, organisation and authority involved in the development and implementation of large-scale construction plans should aim at sustainable solutions to avoid adverse impacts and conflicts. The concept of "sustainable development" has made it possible to identify new goals and to offer new opportunities.

What is sustainable development? The classic definition as formulated by the Brundtland Commission is: "Development that meets the needs of the present without compromising the ability of future generations to meet their own needs". While accepting the original definition as an intellectual, practical and political challenge, we might reformulate the definition as follows: sustainability is not a fixed state of harmony, but rather a process of change in which the exploitation of resources, the direction of investment, the orientation of technological development and institutional change are constantly adjusted to future as well as present needs.

A sustainable society does not need to be stagnant, boring, fixed or unadaptive. It does not require rigid or central control, or to be uniform, universal or undemocratic. It could be a world of innovation and development without growing beyond its limits. The transition to a sustainable society requires more than just productivity and technology, it also requires maturity, compassion and wisdom.

Sustainable solutions concerning economic development have to integrate environmental concerns and to combine two main objectives: 1) to secure human values; and 2) to respect the carrying capacity of the affected physical and biological systems. This means that we must observe principles of fairness in assessing the value of nature. Modern technologies allow us to live in a closer harmony with our environment than we have been able even in the recent past. Business as usual is not a sustainable option. There should be a greater respect for nature in terms of its beauty, uniqueness and ecological integrity. Protected areas and human environmental concerns in particular should be respected. Protected areas have important functions. As formulated by UNEP 1991: "Natural resources can be turned into cash, but cash cannot be returned to natural resources".

These overriding aims have to be transferred into management. In their dealings with each other, stakeholders should return to rational discussion. Principles of environmental ethics must become a matter of the highest priority. In dealing with environmental issues codes of moral conduct and practise should be compiled and implemented as soon as possible.

3. Integrated management approaches demand an open participatory process with the involvement of all stakeholders. Consequently, all involved parties must be taken seriously. Only this will lead to greater readiness for mutual respect and cooperation. It is necessary that the stakeholders are involved from the early planning stages. By this, a better understanding of opposing points of view and the concern for feelings or worries about uncertainties will be promoted.

It is of crucial importance that everybody is treated equally and has access to all the information required for critical evaluation of the activities, their ongoing development and their ultimate consequences. In this way stakeholders will show more willingness to seek consensus on what to consider as acceptable impacts and appropriate compensation measures.

The whole process of project planning and implementation is becoming a joint effort to find solutions which are acceptable to everybody. Taking the matter one step further, it is suggested that the public must be involved as well. The acceptance of large-scale constructions with potentially adverse effects on the physical, biological and social environment will be much better if the public is continuously informed about the different phases of project implementation. Public hearings must therefore take place. People will identify with a project and ongoing activities if they are pro-actively involved. A good example are the windmill parks in Denmark. Local people have a share in the parks and thus consider the windmills as their own. This is a completely different strategy from that adopted in other countries and, as a result, the acceptance is much higher. Since the public has to live with the changes caused by constructions, a pro-active involvement avoids unnecessary conflicts.

Despite all good intentions there will be cases where even an open dialogue between stakeholders will not result in consensus because of extreme differences of opinion. In such cases independent mediation should be made available to maintain a proper balance of interests and viewpoints and to ensure that basic ethical principles are observed. It should be emphasised that the mediation must be fully independent in order to have the trust and acceptance of all stakeholders and the public in general.

In the recent past, involved authorities very often tried to function as mediators. In most cases such attempts went wrong. Stakeholders and other interest groups were wary of such action because the mediating authority was a stakeholder, decision maker and mediator at the same time. Thus, independent mediation is a good instrument for environmental and social conflict resolution.

4. Integrated management needs a firm basis in legislation to provide long-term planning reliability. At present, responsibilities are spread over too many authorities and it is often difficult to come to integrated solutions. As mentioned above, project fora could concentrate responsibilities. However, national legislation is in dire need of reform and reduction in bureaucracy practically world-wide.

Ideally, environmental policy should be considered as a socio-economic service with well defined objectives in order to ensure the development of sound management policies. Integrated management means greater cooperation between the decision makers and the stakeholders. Instead of formulating procedures and applications for everything (which is the responsibility of the companies), political decision makers should have the responsibility of supporting and implementing integrated solutions formulated in project fora.

The requirements of guidelines for integrated management have already been mentioned above in connection with the integrated approaches. Additionally, environmental costs must be taken into account in order to include the value of nature. The ambitious aim of managing future developments in a sustainable way needs a far-sighted comprehension of the rapidly changing world.

Integrated management stands out by including economic, ecological, social and ethical requirements in the decision-making process in the course of project execution. There are other basic guidelines of environmental management systems which are part of holistic, integrated management; for example:

- Open-mindedness for innovation must be actively promoted.
- Top management and project leaders need to broaden their intellectual horizons.
- Employees must be trained for an open participatory process.
- Teamwork must be made possible, as it is an essential requirement for cooperation.
- Progressive attitudes of companies and of employees must be promoted and supported.
- Additional responsibilities in several fields of work have to be created.
- Shorter routes of information and communication must be developed.

On the basis of such guidelines, sound management policies can be formulated for more effective project execution.

5. Today's problems concerning economic development are becoming increasingly more complex and sometimes even more chaotic, thereby gradually exceeding the capacity of present political systems. Therefore, environmental policy and management procedures should take into account that there is a growing need to operationalise the precautionary principle and not to simply adopt an "end of the pipe strategy".

We have to distinguish between corporate environmental policy and precautionary principles defined by the state. Corporate management policy should actively formulate and cover environmental guidelines and performance criteria. Companies should demonstrate an environmental responsibility, and should develop a detailed environmental protection plan (EPP) for every stage of planning, construction and operation of a project. Companies need to be more visible by demonstrating positive environmental action and by subscribing to international guidelines and codes of conduct.

Environmental policy by state is based on three main principles: 1) the polluter pays principle; 2) the cooperation principle (integrated approaches); 3) the pollution prevention principle.

The latter should be extended to include as an important precautionary principle the avoidance of negative impacts on physical habitats. Temporary impacts in most cases cause short-term effects, but there are others which cause long-term effects to the extent of destroying physical habitats and their ecosystems. The precautionary principle has to evaluate very carefully the implementation of large-scale projects to avoid detrimental impacts as far as possible. There is an urgent need to take account of regional differences and existing protected areas since protected and other sensitive areas have important environmental functions.

In the development of improved environmental policies and management aims, science has an important role to play. However, scientists should improve the publicising of their contributions, while communication between scientists and politicians must also be improved. Therefore, scientists must learn to make clear and understandable statements without the use of scientific jargon and decision/policy makers should learn to formulate clear questions.

A possible contribution of science is the establishment of effective monitoring programs to support decision-making processes by providing objective data and facts. By means of monitoring programs we obtain the necessary data base and improve our understanding of the function of ecosystems and their natural and man-made changes. In this way it is possible to identify the impacts on the environment. Furthermore, the assumptions made in EIAs can be checked and project evaluation is made possible.

6. The proliferation of large-scale constructions in sensitive areas requires application of the precautionary principle and the consideration of the value of nature. The method of cost benefit analysis (CBA) is a good tool for this purpose. Standard CBA in general is a method of calculation of economic efficiency, but since environmental resources and nature protection are inseparably linked with economic development, CBAs should include all environmental, social, cultural and transborder-derived effects. Market economy may ruin the environment, and hence ultimately ruin itself, if the price structure does not reflect the ecological value of the environment. This aspect has been totally ignored in the past, but today its significance is increasingly being recognized.

Wherever feasible and practicable, compensation for the loss of or the damage to environmental resources and natural assets should be charged to the economic project. These compensation costs would then be added to the development and project implementation costs.

Furthermore, these costs must be included in the cost benefit calculation. Magnitude and significance of environmental impacts need to be standardised in monetary terms. These methods have to be transferred to social and cultural values to include all secondary effects of economic development.

CBA could have great power as an heuristic method to an environmental and holistic management approach. It should observe equity principles in the process by including monetary valuation methods and techniques in assessing the environmental effects. In the context of managing large-scale projects in the most sustainable way, the application of the best available technology (BAT) should be considered. However, its application must be carefully balanced to take account of regional and cultural differences. Furthermore, BAT should not be used uncritically but should rather be justified by specific measurable environmental benefits.

7. Environmental impact assessment (EIA) plays an important role in project implementation. It gives the environment its due place in the decision making process by evaluating adverse impacts on the environment before any construction activities commence. The UN conference on Environment and Development in 1992 recommended the global application of EIAs: "Environmental impact assessment, as a national instrument, shall be undertaken for proposed activities that are likely to have a significant adverse impact on the environment and are subject to a decision of a competent national authority".

While there are some guidelines and objectives available for EIAs, they should nevertheless be constantly improved by learning 1) through adaptive application during construction and 2) through critical evaluation after project completion. However, EIAs need a clear legislative basis and regulations for transboundary control are required. Large-scale constructions very often cause transboundary effects, as in the case of water reservoirs or fixed link projects between countries. Besides bilateral regulations, universal and international guidelines are needed.

Furthermore, EIAs have to be initiated before the beginning of project planning and have to be integrated into the project development. Industry itself should use this instrument more actively during planning, construction and operation. This improves the project from both environmental and economic points of view. In addition, assumptions and predictions of EIAs should be critically reviewed during construction for the purpose of taking immediate remedial action.

Beside the legislative basis, EIAs need another supporting leg in the form of an adequate data base. The data base is also an essential requirement for the design of environmental impact studies (EISs). Only competent multi-disciplinary teams applying interdisciplinary principles will guarantee a successful EIA. This includes the capacity for quantitative predictions and the recognition of feedback loops. Another essential ingredient are the monitoring programs mentioned above. Project and post-project monitoring will improve the evaluation of EIAs and thereby support an environmental auditing.

Sequential EIAs and monitoring programs are good management tools and should therefore always be linked. The integration of these instruments leads to more realistic formulations of critical questions concerning the effects of project implementation. Research activities are focused on essential aspects and feedback for the assessment of the assumptions of the EIAs is much easier and clearer. In practise

EIAs have proved to be among the most successful approaches for making decisions about policies, plans, programs and projects.

8. Integrated management for sustainable development needs broad knowledge and acute sensitivity. Modern management is increasingly becoming interdisciplinary and more flexible. However, interdisciplinarity is truly difficult to put into effect because there are many different cultures of knowledge and understanding. Achieving integrated approaches in project development therefore requires close collaboration between the policy makers, economic managers and scientists of the various cultures involved.

Every discipline has its own technical language which is usually not understood by the other disciplines. Communication programs are therefore needed to translate the jargon of one stakeholder for the others. In addition, both scientists and decision makers must learn to make clear statements and to formulate clear questions.

Education should be seen as an opportunity to meet the needs of interdisciplinary action and not merely as an unavoidable inconvenience. Many traditional educational institutions are currently unable to offer interdisciplinary training programs. In many cases appropriate training programs still have to be developed and the educational facilities to be made available.

Interdisciplinary approaches for a sustainable development in the future are more than simply an integration of different attitudes or theories. A completely new style of management is required, one which is more consultative, pro-active and participatory. This should be a flexible process as regarded by some of the old Greek philosophers who recognised that everything is in a state of flux. The process of learning and communication is a challenge to meet the needs of the future.

In the preamble of the recommendations it is pointed out that the increasing number of large-scale constructions exert strong pressure on the environment. In this context we need a modern educational philosophy linking the environment with development. The conference showed that sensitivity, and especially the time scales of the environmental system, as well as social and economic impacts are poorly understood. Though political conditions are often short-sighted, transborder government programs have to supply the necessary educational facilities. This means that both industry and government must make an effort to guarantee a satisfactory development of further education. The UN should support training courses in integrated management designed to bring together industrial and scientific partners in order to improve competence in fields such as: 1) holistic strategy resource use of the environment; 2) environmental communication; 3) integrated management with a focus on social and economic systems; and 4) education in ecosystem functioning.

Acknowledgements

As Conference Organiser, I would like to thank the members of the Steering Committee for the many fruitful discussions and for their most helpful contributions during the planning, development and the organisation of the First Symposium on Large-Scale Constructions in Coastal Environments. Furthermore, I am indebted to the conference sponsors, particularly Statoil Deutschland GmbH, for their support in making the symposium and this publication a success. I am grateful to Wim Salomons for making publication possible and, finally, I would like to express my thanks to Sabine Muslimat and Dirk Saretzki for their assistance and encouragement during the conference on Norderney.

Manfred Vollmer

Editors

Henning Grann
Statoil Deutschland GmbH, Jannes-Ohling-Straße 40, D-26723 Emden, Germany

Manfred Vollmer
ENCORE, Weserstraße 78, D-26382 Wilhelmshaven, Germany

Steering Committee

Manfred Vollmer	Conference Organiser and Chairperson, Bureau ENCORE, Wilhelmshaven
Luitzen Bijlsma	Rijkswaterstaat, Den Haag
Hubert Farke	Nationalparkverwaltung Nieders. Wattenmeer, Wilhelmshaven
Burghard W. Flemming	Senckenberg Institute, Wilhelmshaven
Henning Grann	Statoil Deutschland GmbH, Emden
Henning Karup	Danish Environmental Protection Agency, Copenhagen
Rainer Knust	Alfred Wegener Institute, Bremerhaven
Helge Rakstang	ENS-Foundation, Stavanger
Wim Salomons	GKSS Forschungszentrum, Geesthacht
Karlo van Bernem	GKSS Forschungszentrum, Geesthacht

Paper Selection and Editing Panel

Per Albricktson	Statoil Norway, Stavanger
Alexander Bartholomä	Senckenberg Institute, Wilhelmshaven
Luitzen Bijlsma	Coastal Zone Management Centre, The Hague
Øystein Dahle	Den Norske Turistforening, Oslo
Monique Delafontaine	Senckenberg Institute, Wilhelmshaven
Sabine Dittmann	Terramare Research Centre, Wilhelmshaven
Wolfram Elsner	University of Bremen, Bremen
Jens Enemark	Common Wadden Sea Secretariat, Wilhelmshaven
Birgit Georgi	Umweltbundesamt, Berlin
Henning Grann	Statoil Deutschland GmbH, Emden
Frank Krögel	Varel, Germany
Vincent May	Bournemouth University, Poole, Dorset
Bastian Schuchardt	BioConsult, Bremen
Manfred Vollmer	ENCORE, Wilhelmshaven

Conference Sponsors

Statoil Deutschland GmbH
Directoraat-Generaal Rijkswaterstaat
GKSS Research Centre
Bezirksregierung Weser-Ems

Conference Support

UNEP, United Nations Environmental Programme
UNECE, United Nations Economic Commission for Europe
PIANC, Permanent International Association of Navigation Congresses
WWF, World Wide Fund for Nature
ENS, Environment Northern Seas Foundation
CWSS, Common Wadden Sea Secretariat
BMBF, Bundesministerium für Bildung, Wissenschaft, Forschung und Technologie

Preface

On 21–25 April, 1997 the First International Symposium on Large-Scale Constructions in Coastal Environments was held on the island of Norderney in Germany. The Wadden Sea National Park Authorities in Lower Saxony initiated this symposium which was inspired by the experience gained through the planning and implementation of the Norwegian State Oil Company, Statoil, natural gas pipeline landfall project, Europipe, during the period 1991–1995. The overall objective of the symposium was to provide contributions to the resolution of the traditional conflict between the need for economic development and the concerns for protection of the environment.

During the early planning, the scope of the symposium was extended to include two other major coastal construction projects, in addition to the Europipe project, namely the Danish/Swedish fixed link projects and the Dutch coastal protection projects. Furthermore, an invitation was issued to potential participants to present similar case studies or topics which would stimulate discussion within the areas of environmental policy, economy, ethics or science.

The overall objective was achieved through the development of a number of specific recommendations prepared by all symposium participants in a joint effort. Several recommendations are of general application and thus not limited to coastal construction projects. The main thrust of the recommendations was to stress the need for development of integrated approaches to project planning and implementation. The importance of an open participatory process aimed at sustainable solutions with due respect for the value of nature and the observance of ethical principles was underlined. Cooperation between stakeholders rather than confrontation and conflict was highlighted as a necessary prerequisite.

The recommendations were officially handed over to representatives from the Lower Saxony district government of Weser-Ems and from the three organisations, United Nations Economic Commission for Europe, United Nations Environmental Program and Environment Northern Seas Foundation, for their consideration and subsequent submission to national governments and various nongovernmental organisations as appropriate.

This volume contains a number of articles, most of which were presented at the symposium and considered of interest in future efforts to achieve sustainable project solutions. Also included is the full text of the conclusions and recommendations developed during the Norderney symposium.

Contents

List of Contributors

Dr. Asokkumar Bhattacharya
Calcutta University
Dept. of Marine Science
35, Ballygunge Circular Rd.
700 019 Calcutta, India

Dr. Andreas Brenner
Universität Potsdam
Politische Theorie
Postfach 90 03 27
D-14439 Potsdam, Germany

T. Neville Burt
HR Wallingford Ltd.
Howbery Park, Wallingford
Oxon, OX10 8BA, UK

Ian Cruickshank
HR Wallingford Ltd.
Howbery Park, Wallingford
Oxon, OX10 8BA, UK

Karen Edelvang
Danish Hydraulic Institute
Agern Allé 5
DK-2970 Hørsholm, Danmark

Steinar Eldøy
Statoil HMS & K-T
P.O. BOX
N-4035 Stavanger, Norway

Prof. Burghard Flemming
Senckenberg Institut
Schleusenstr. 39a
D-26382 Wilhelmshaven, Germany

Henning Grann
Statoil Deutschland
Jannes-Ohling-Str. 40
D-26723 Emden, Germany

Dr. Carolyne Heeps
Bournemouth University
Talbot Campus, Fern Barrow
Poole Dorset BH12 5BB, UK

Prof. Thomas Höpner
Universität Oldenburg
ICBM
Carl-von-Ossietzky-Str. 9–11
D-26111 Oldenburg, Germany

Henning Karup
Danish Environmental Protection Agency
Strandgrade 29
DK-1401 Copenhagen K, Danmark

Jon Larsen
KSÖ
Länsstyrelsen
S-20515 Malmö, Sweden

Mervyn A. Littlewood
HR Wallingford Ltd.
Howbery Park, Wallingford
Oxon OX10 8BA, UK

Prof. Vincent May
Bournemouth University
School of Conservation Sciences
Talbot Campus Fern Barrow
Poole Dorset BH12 5BB, UK

Monika Puch
KSÖ
Länsstyrelsen
S-20515 Malmö, Sweden

Ralf Röchert
WWF Deutschland
Fachbereich Meere und Küsten
Am Güthpol 11
D-28757 Bremen, Germany

Dr. Bastian Schuchardt
BioConsult
Lesumstr. 10
D-28759 Bremen, Germany

Christina von Schweinichen
UN Economic Commission for Europe
Human Settlement Section
Palais des Nations
CH-1211 Geneve 10, Switzerland

Dr. Aad C. Smaal
RIVO-DLO
P.O. Box 77
NL-4400 AB Yerseke, Netherlands

Prof. Kerry Turner
University of East Anglia
School of Environmental Science
Norwich, Norfolk NR4 7 TJ, UK

Alle van der Hoek
Rijkswaterstaat
P.O. Box 20906
NL-2500 EX The Hague, Netherlands

Wouter van Dieren
IMSA Amsterdam B.V.
Van Eeghenstraat 77
NL-1071 EX Amsterdam, Netherlands

Sue Wells
WWF International
World Conservation Centre,
Av. du Mont Blanc
CH-1196 Gland, Switzerland

Part I
Environmental Ethics

CHAPTER 1

The Tower of Babel was a Coastal Edifice

Wouter van Dieren

The subject of this chapter is, to the outside observer, not a very dramatic one. Clearly, experts from along the coasts of northern Europe get together and tell each other stories about the technical miracles they perform. Their technology is, probably, some 5000 years old, or even more. Wherever there is a coast, a river delta and a resourceful upstream hinterland, there is trade and other economic activity, and on these spots cities were built and harbours constructed. This is all history, and it will be future, *Deo volente*. The question is what the environmental commentator would tell you at this very occasion. No doubt you discharge your duties with care and respect for the natural beauty of these coasts, respect for the rare, sometimes endangered animals which live on the sands, the flats, the tidal estuaries, the dunes and the pastures of these blessed lands. It takes very little to disturb it, but the effort to preserve it is heavy and bears great responsibilities for all those involved.

Yet, it is with some fear, and also diligence, that I observe your activities. The sturdiness of coastal engineering, the powers of the interventions, and the sheer optimism of the technician do create a sense of alarm for the naturalist, because these are the signs of danger. Notwithstanding the influences of the green wave in the last decades, we observe a victorious revival of the old economic order, which believes in expansionist economies and eternal economic growth. There is little doubt that the ensuing energies will cast dark shadows over our coasts.

This chapter therefore deals with the tensions between desired next "coastal greatnesses" and the need for extra care and prudence. Assuming that the world is at the cross-roads of a next, final blow at nature or a diversion into a world which we call sustainable, I will address you with a plea for the latter; to ensure that the coast remains what it was and is – the transformation zone between water and land, the edge where one day oceanic life crawled onto the land. It is here where nature is most vulnerable and yet most dynamic, where all life began, but also where it can end.

There is no doubt that man is able to design very large structures which mimic the Creation itself. More over, this is what he really wants, though in the knowledge that this is an act of profound blasphemy. Throughout history, any attempt to depict the natural world was considered an act against God Himself; He being the Only Creator, who could make forms out of Nothingness, life out of clay. For man to do likewise was a trespass on the divine prerogative, and, of necessity, a parody of it.

The most impressive of all these parodies, and the most punished, is in no doubt the Tower of Babel, which in the three paintings by Pieter Breughel the Elder is a very large-scale coastal construction. The Tower and its history represent the ultimate metaphor of vanity, arrogance and ignorance, where man tried to become equal to God, in efforts to get access to His throne. Also, the Tower is the earlist known metaphor for the impossibility of continued economic growth on a limited planet.

In the vision of Breughel, the Tower is an edifice, erected on a European coast instead of in a Mesopotamian desert. After a hundred years of strenuous construction, the Babylonians still had not caught a glimpse of the throne of God, so the conclusion had to be that Heaven was to be found a few miles higher than where the summit of the Tower was at that time. Whatever they tried, though, they could not add a single foot to the top, and worse, they witnessed to their surprise a gradual dismantling, such as the crumbling of arches and maintenance backlashes of the serpentine road around the structure. Resources had to come from ever larger distances, and they were ever more needed for the transport itself of these resources and for maintenance purposes underway. Yet, the technicians did not dare tell the ruler of Babel that he would never reach his target, and the economists kept telling him that here was still economic growth, measured by the number of ships arriving in port, and the number of donkey and oxen carts departing for the long voyage to the top – where they would never arrive. The ensuing confusion was total, in the Babel legend itself expressed by means of the penalty of the many tongues, so people could no longer understand each other. Thus consequently, Babylonian civilisation came to an end, and the Tower collapsed.

Technology and wealth can provide a good symbiosis, but they can also do the contrary, as this story makes so explicit. Glorious civilisations can flourish with the help of technology, yet most of them have vanished, after events which throughout history have resembled the fate of the Babylonian Tower. However, the arrogance of man is beyond measure, and time and again he restarts his adventurous efforts to change the fate of the earth without respecting its limits.

There is no doubt in my mind that we are indeed facing a similar course of history now. In spite of the upsurge of many Cassandras, which warn against evil and danger, or whose sisters on the eleventh floor of the Tower send messages down below that growth is over (and in spite of modern confirming statistics), the world is still not capable of proper governance and common sense, of self-control and resistance to greed. In the wake of the next millennium, we are now subjected to global competitive economic warfare, to neoliberal vanguardism, to profit for the few and misery for the many.

For coastal development, the story is either grim or transparent. Grim it is if we want more of the same, if we indeed plan for additional ugliness, for the aggressive structures of the technological past. Any coastline can become a next Rotterdam, or a Belgian casino coast, or a Côte d' Azur (which is in reality an experience of bad taste and polluted air). It is not difficult to conceive of a harbour line all the way from Rotterdam to Aarhus, only interrupted by nouveau riche developments as on Sylt, Borkum or in Zandvoort.

However, the requirements for sustainability are very different from those traditional interceptions. The real innovative powers of the future will have to build on a wholly different baseline. If the future is to be easy, in that it is to be a replica of the present, then I would not be standing here. If the future has to be different, for compelling reasons, then we need imagination and the courage to change course.

There is no doubt that the prerequisites of the future are very different from these conditions which prevail over today's practices. We refer to these differences as discontinuities. They differ in urgency and nature, but it is their combination which makes the difference. Below (see Sects. 1.1–1.10) we discuss a few.

1.1
Taking Nature into Account

Environmental costs have to be internalised. Scarce goods (like raw materials and energy) must have a higher price than relatively abundant ones. As a result, raw materials will have to be used sparingly and prevention of environmental damage financially stimulated. The costs are to be internalised by means of taxes, subsidies, tradeable rights, and laws and regulations.

1.2
A Drastic Increase in Productivity

Since industry and authorities will have to internalise environmental costs (see Sect. 1.1), they shall and will strive to improve the efficiency of raw materials and energy, instead of the old focus on labour efficiency. In this field a true revolution will take place (the efficiency will increase by a factor of 4 in 20 years and a factor of 10 in 50 years), which will affect product design, production technology and consumption (dematerialisation). It has to become common policy to minimise the material input per unit service (MIPS).

1.3
New Policy Indicators Measure Welfare

The old emphasis on income growth has to be replaced by a broader definition of welfare as a criterion for progress (environment, social aspects, health, material welfare). New policy indicators (national as well as regional) will provide insight in to this broad view of welfare. Within the next 10 years, globally, the GNP will have to be adjusted for environmental losses, and so the green GNP is coming into sight.

1.4
Concentration and Separation of Spatial Functions

Spatial functions (like buildings, recreation, nature and agriculture) will have to be geographically strictly demarcated. Areas must have a clear function and clear boundaries. Large contrasts will have to be recognised by gradual transitions, like water-purifying swamps or park-like recreation areas between city and nature.

1.5
Explicit Management/Conservation and Development
of National Parks

Large nature reserves will have to be safeguarded, and their self-regulation restored. Green and ecological functions must be fully developed. In natural areas the infrastructure will have to be reduced and constructed according to the latest technical concepts with the least possible damage to the area.

1.6
Material Flow Chains are Closed Where this is Environmentally Most Effective

Cities of the twenty first century cities must thrive on sustainable energy (in anticipation of depletion and change of resources, in combination with a high degree of energy efficiency). Water will have to be supplied in two qualities. Rainwater will have to be drained away separately and re-used. Swamps must serve to purify water in a natural way.

1.7
Use of Energy-Extensive and Non-Toxic Materials in the Construction Industry

Health will tbecome the central objective in architecture, meaning that the materials used must be non-toxic, free of radiation and save energy and raw materials. The basis for building construction will have to be timber instead of concrete.

1.8
Environmental Planning as an Instrument to Minimise Transport

Proximity instead of good transport must be the central objective for future planning and, as such, the key principle to determine locations: housing and jobs will have to be combined (in urban areas the main residential areas and industries will have to be located within a radius of 8 km); companies must be linked up with public transport.

1.9
Minimal Dependence on Fossil Energy in Transport

Energy-extensive forms of transport must be stimulated. In the transition phase to electric transport or transport using hydrogen, the (urban) public transport will more and more use biodiesel. The energy use and emissions of cars and trucks will be drastically reduced. The hyper-car is in design, being an all-fiber vehicle of light weight and absolute minimum oil use ($2.1 \, l \times 100 \, km^{-1}$). Public transport will be cheap, frequent, comfortable, optimally interlinked and on call, and combined with communication technology. The light-weight vehicle is rapidly developing.

1.10
Minimising the Transport of Goods

Dematerialisation and changed production structures will drastically decrease the need to transport goods. Proximity as a criterion for the location of companies equally decreases the transport need. The availability of goods will be guaranteed at any time by means of an optimal logistic system with the smallest material input per unit of product or service. This system will be backed up by information technology for supply and demand (also between production chains), leading to improved efficiency. Tube transport instead of lorry (LKW) will be widely used, providing a factor of 25 MIPS increase.

1.11
Final Remarks

For the coast, the difference is in the technology and the design. In a sustainable world, there is no next destructiveness, no next conversion of nature into an ugly port, where ugly goods are being transloaded from ugly container ships into ugly trucks and lorries, that carry them to an ever more uglier hinterland, to be there consumed and disposed of on ever larger and uglier waste mountains. This is all we have done for a century or so, and one day it must stop. There is no future in further destruction.

A next coastal development in The Netherlands, therefore, has to be a definite journey into the age of beauty. So be it that sea levels will rise, due to global warming – then though the Dutch shall not construct another huge dam of steel and concrete! On the contrary, the next coast inevitably has to be built, but it has to become a coast of nature, where all we have learned about ecosystems and natural reserves find a place, the final natural harmony between man and technolgy. For the Zuiderzee, we made such a design, which is comparable. Other than the known tradition of heavy dikes and mathematical structures to effectively block the sea, we designed a flexible landscape, the Almeria Archipelago, a wilderness of inlets, islands, swamps, wetlands and waterways, where almost all natural functions will benefit. It will end, for instance, the onslaught on birds, it will improve birdlife, fishlife and the diversity of plants, it will clear the water quality, and it will create surprise and adventure in a country which is now so void of natural resources. A next North Sea coast should be no different; a flexible coast, not a sturdy one. A coast full of inlets, lagoons and new estuaries. The black stork will live here again, as well the white heron, but also the dolphin and the smaller whales should return.

The future is not about a repeat of the past. The future is not about Big is Beautiful. The future is about care and prudence, about integration and harmony with the living world.

CHAPTER 2

Philosophy and Man's Responsibility for Nature

Andreas Brenner

2.1
Introduction

1997 has been a year of important anniversaries for environmental politics: 5 years for the Conference of Rio, 10 years for the Conference of Montreal and 10 years for the Brundtland Report *Our Common Future*. For politics these anniversaries should be a cause not only to declare new political intention but also to critically reconsider what has been achieved so far. Politics has to answer the question of why it was not more committed to saving nature. Why did the ozone layer have to be destroyed and the risk of climate change to be taken before politicians realized there was a big problem to solve? Why does it take such a long time for a paradigm shift? It is my thesis that this paradigm shift in politics cannot emerge before philosphy by itself develops a new paradigm. The responsibility of philosophy, however, is greater and extended to the structure of thinking and acting which causes the destruction of nature.

In order to examine the relation between philosophy and man's responsibility for nature the first thing that has to be discussed is the role of traditional philosophy, both in general and in the development of modern natural sciences. The precondition for the rapid development of natural sciences was a change in the way man perceived the physical world. This resulted in a change of conception of the world and of the human being as such. The beginning of this process can be observed in the transition from the Renaissance world to modern times. Two individuals were of particular importance in this dramatic period: Leonardo da Vinci and Francis Bacon.

Leonardo's (1452–1519) profession as a painter led him to explore the human body. He had already painted his most famous pictures, *The Last Supper* and the *Mona Lisa*, when he began to investigate the human body. However, with simple observation he could not satisfy his curiosity. Leonardo felt that he had to search deeper and started to dissect human bodies, in total more than 30 human corpses. The results of these investigations have been preserved in his drawings. The goal of Leonardo's dissections was to learn how nature worked. Through his new knowledge he hoped he would be able to paint better pictures. However, Leonardo was not only a painter but also a great engineer (Maschat 1989). He based the constructions of machines on his study of biological mechanisms. When, for example, designing an aeroplane he tried to copy a bird. In this way Leonardo stands for the interest and respect with which Renaissance man regarded nature.

Francis Bacon's (1561–1626) principles of investigation were completely different from Leonardo's. Whereas Leonardo's studies were the result of pure curiosity aiming at knowledge for its own sake, Francis Bacon did not consider knowledge as a goal in itself but as a stepping-stone in a process of application. In order to ensure this outcome Bacon needed a method which would make such investigations uniform and therefore reproducible. Experimental investigation should be superior to that of an artistic contemplation of nature. The experiment should be independent of who performed it, i.e. objective and universally valid. He wanted the observation of nature to be carried out with the goal of applying the knowledge. In order to achieve this goal Bacon advised people to literally keep nature on tenterhooks. Bacon had been a lawyer by profession and had been lord-chancellor before he started his scientific investigations. He believed that torture was a way of finding hidden truth and that physical manipulation of nature would reveal its secrets. This concept resulted in nature losing its aura and becoming a purely material substance (Gloy 1995).

2.2
Modern Science and Modern Philosophy

In modern times the loss of nature's aura was not recognized as such, because the advantages of the new scientific method were increasingly shared by all. This is evidenced by the modern world which surrounds us by comforts produced by science and technology. The health service, the technology of transport and of communication make life more comfortable for all.

As a result man has emancipated himself from the domination of nature as well as from ancient and oppressive social structures. Hence, modern society is based on two concepts: 1) modern science guarantees man's emancipation from the arbitrary power of nature which is paralleled in modern social philosophy where man has been liberated from arbitrary political power. Thus, an attack on one has often been seen as an attack on the other. So if we reinstate the sovereign position of nature, may we not also be in danger of subjecting ourselves to some new form of political tyranny? In other words, the modern view of nature becomes sacrosanct because an attack on its basic principles could be seen as an attack on the principles of human freedom as expressed in the Enlightenment. 2) on the other hand, there is no doubt that our present exploitation of nature is destroying man's future. Therefore we have to find a new relation to nature not opposed to modern society but integrated into modern society. In order to realize this project the term autonomy becomes important.

Our question is, how can we respect nature in a modern society? Modern society, as a result of the Enlightenment, is characterized by a high level of self-determination. The Enlightenment established the modern social subject who freed himself in two ways: the first act of liberation occured in the field of knowledge and the second took place in the field of society and politics.

Concerning the first act, we find the most important references in the works of Decartes and Kant. Descartes' famous dualism divides the world's phenomena into two different kinds: material and mental phenomena. This division became a basis for a better orientation as well as a new insecurity: on the one hand people feel more insecure, because they see the division of the world as part of their own personality. Therefore, they are in danger of not conceiving themselves as a unity any more. People have to face the question: Who am I? What is the real nature of my identity? On the other hand this feeling of uncertainity can be compensated by the idea that nature as such is the outside, nature is the other and is not protected by law.

Four hundred years after Descartes' birth we can look at the history of the Cartesian dualism. What has been achieved? Today we live in a world of endless possibilities.

Almost whatever we imagine we can realize. Every dream of Jules Verne and further-more other dreams of mankind have been realized. The flight to the moon is no longer a utopia, the voyage around the earth does not take 80 days but less than 24 h and nature is no longer man's enemy but man's slave. We are the children of Bacon and Descartes and our epoche is characterized both by the realization of utopia and by the end of utopia. In this respect dreams of mankind have become true while we are on the other hand missing imagination and creativity to cope with the dark sides of these dreams; the list of our problems is long: the hole in the ozone layer, the waldsterben, the endangered species of plants and of animals and so on.

2.3
Philosophy's Responsibility

If we consider the previous thoughts, we cannot deny that the actual development of society, technology and sience has been initiated by philosophy. By the same token one could establish the hypothesis that without occidental philosophy we would not be con-fronted with the greatest danger in the history of mankind: the imminent collapse of the earth's ecosystem, or the ecological crisis.

In this situation we also have to ask whether it is rational to have confidence once again in philosophy. Is it an illusion to expect any useful idea from philosophy? Phi-losophy has a rich tradition of asking questions related to the meaning of human acts. It is in this way that philosophy may contribute to the understanding of our acts and to their integration into a rational connection between the ego and the other, between the individual and the community. This competence philosophy has developed in the course of the past two and a half millenia. Beginning with Aristotle the ethics ask for the con-ditions of a good social existence. Since then ethics have been looking for rules which orientate the human relationships to criteria other than physical, intellectual or politi-cal power.

It took over two millennia for philosophical knowledge to change the field of society and politics. The most famous documents of this success are the new constitutions of the USA (1787) and of the French Republic (1789). There was established a political foundation for modern society. Today nearly all states of the world have accepted the thoughts of equality and of human rights. This fact is quite important when you see that political practise often works against theoretical achievements. This does not re-duce the importance of the theory manifested in those two constitutions: it is the basis of modern law that gives people the chance to appeal.

Now, we have to ask ourselves whether it is a realistic hope that philosophy can change the relation between man and nature in the same way as it did among men in the past. Or, is it possible to convert the relationship of man and nature according to the analogy of the relationship among men, which is characterized by an equal and non-hierarchi-cal situation. It was the idea of the French Revolution that people should be sisters and brothers. Is philosophy able to revolutionize in a similar way our relationship to na-ture? Can nature be the brother of mankind?

Nature as an equal partner, as a friend or a brother of man, is an old picture in the mystification of nature. Even today you can find parts of this mystification in two fields: in the old archaic societies and in numerous religions. In traditional societies, in the South Seas for example, there is the idea that it is possible for man to communicate with nature. These people have the custom of apologizing to trees before cutting them down.

Such a custom is not found in modern civilizations – it would be incomprehensible in our society. However, in modern societies you find other elements of a deep relationship between nature and man. In this regard modern religions revive the idea that the whole of nature, man, animals and plants have a common creator: God. For the faithful, creation is a gift from God and its destruction is a sin against God.

In our context the idea of creation has a great weakness: it is valid only for religious people. It is evident that people who do not believe in God would not respect a religious law. Anyway, the thought of creation is interpreted in the different religions in a different way. Religions are inevitably particularistic.

In view of the ecological crisis, there is the need for a universal theory, a theory which is entitled to have total validity irrespective of different world views or different interests. This is the criterion of modern philosophy. In the field of ethics it was Immanuel Kant who first introduced the concept of universality: with the categorical imperative Kant gave us a rule for acting which is valid in every place and every time. The categorical imperative gives the order: act in the way that you would want, that the principle governing your acting could be a general law, governing all such actions all the time (Kant 1974). This sentence has a pure logical function. It is the aim of the categorical imperative to avoid such acts which have an internal inconsistency. These are acts which Kant said you cannot want them.

For the classical ethical rules Kant succeeded in proving that it is not possible to act against the rules without involving oneself in a logical contradiction. A person, for example, who lies, does harm not only to other people but also to himself: he constructs a fictional world which depends on the fact that he is part of a minority which profits from the general acceptance of rationality. To lie, to harm people, all of this is possible, as it happens every day, but you cannot find rational reasons for it. Such acts are arbitrary, non-objective and in a philosophical way you can say they remain solitary.

The importance of Kantian ethics is established for micro-ethics, that is the part of ethics that is concerned with the relationship between individuals. Philosophy has not yet sufficiently discussed the relation between man and nature, and between the present generations and the future ones. Are there also macro-rules?

If we discuss ecological problems we have to use the same criterion that we learned from Kant: should we cut down the tropical tree, shall we continue to produce dangerous substances and other questions have to be answered in accordance with the Kantian categorical imperative. So we see that logical consistency, which would provide a criterion, is identical to the idea of having a future: is it possible to extend a special act to the future? Can you imagine making other projects with the same structure in every place and in a large number? This is the criterion we could learn from philosophy. In the ecological debate we know a similar criterion, that of sustainability. Philosophy has to pick up this idea and make it a criterion for action. This means that we have to orientate our action towards the criterion of sustainability. Our action is measured by the following questions: is a concrete action justified by the criterion of sustainability? Is it also consistent if we consider its consequences in the long run?

Then we immediately realize that many actions are not justified: felling the tropical forest and destruction of the ozone layer cannot be sustained in the the long run. However, we live in the present. What could organize our actions in such a way that they are compatible with the long-run view, given this may restrict our present interests?

The important motivation is the following: it should be everybody's first interest to ensure that his or her action is not self-contradictory. Especially in the culture of Enlightenment the condition for actions as a free man is to maintain our autonomy. This is necessary if we are convinced that we are free, sovereign and responsible beings.

To act really autonomously we have to inform ourselves as far as possible about the consequences of our actions. Only by doing so, will they remain free and autonomous. However, then we must base our consideration on the actual knowledge of consequences. People can only act responsibly if they ensure that they have this knowledge and then decide in the perspective of the long-run.

We can see that many actual projects can no longer be regarded as responsible. The idea of sustainability would not only be important for our way of planning and constructing, but also change the whole of our culture: our relationship to nature would no longer be orientated only towards Bacon's principle but also towards Leonardo's principle.

2.4
Nature Politics

However, what would follow from this paradigm shift? The decisions to modify nature have a high level of complexity: it is not enough to look at the interests of everyone involved in the project. It is also necessary 1) to take into account expanded horizons in space and time and 2) to respect the value of nature. Large-scale productions are characterized by effects in a large spatial and temporal zones: a nuclear power station, for example, supplies a large region with energy. However, in the case of an accident the whole of mankind is endangered, because the clean-up of radiation is a problem for many generations. Another example is the oil exploitation in the North Sea: on one side we need the energy supply, and on the other side we clearly see the jeopardizing of nature. So what is the criterion for the decision?

First you have to distinguish two kinds of interests: those of the present generation and those of the future generations. The interests of today are an example for classical political conflict. Politics has to find methods to reach a consensus. It is the difficulty of such a consensus that the typical case for solving the conflict, the compromise, does not work. What could the compromise achieve if one side wants to construct the power station and the other side wants to prevent it? For the environmentalists a compromise is unacceptable. Their aim is to prevent the construction and neither a reduction nor a postponement can be satisfactory. This, however, is the only way for the industry to accommodate the environmentalists if it plans to carry out the project. Obviously there is no chance of a political conflict solution between both actors. Thus, the conflict will be solved with power. So what would be a political solution?

To achieve a political solution we would have to widen the circle of the participants beyond the directly affected people. It only becomes a political issue if it is put in the context of the circumstances. In this way we have to organize a forum to discuss society's aims and values. In such a discussion the material as well as the immaterial wishes should be reviewed. This would give us the chance to explain our conceptions of nature and to demand respect for nature. Communication of our aims and wishes would be helpful to determine the idea of a good life. In spite of individual differences, a consensus about "good life" would be crystallized. Since Aristotle, the term "good life" characterized a consistent life (Ackrill 1974), or in a modern term a life which is sustainable. Life which is consistent in this way should integrate nature into both personal action and the world (Arendt 1960). This integration of nature must not only exist through a mental revolution: it must also have practical political consequences. One could be to implement the conservation of nature in the constitution (Welan 1984). Next to the constitutional law, nature could have greater political importance if policy were based on the voluntary action of the people. A helpful instrument for such a policy is an ecological plan which was already suggested at Rio (Agenda 21).

The idea of nature's integration would establish a new paradigm of politics (Weizsäcker 1989; Gore 1992). This kind of politics is demanded by the *principle of responsibility* (Jonas 1979). In our actions we have to respect, along with the interests of future generations, the value of nature. This appreciation is an argument of physiocentric ethics. Additionally, another aspect of political philosophy becomes important. The ecological crisis clarifies not only the lack of integration of nature but also the lack of integration of society (Meyer-Abich 1997). More and more often people have to leave their homes because with the destruction of nature their basis of life has vanished (Kane 1995; Sachs 1996). Like war refugees ecological refugees are disintegrated in their own societies. This lack of integration can be found in many countries with terrible natural disasters, e.g. in the Ukraine or in Belarus. The destruction of nature often has winners, but it certainly always has losers. People who become sick from environmental damages are often excluded from society. People who suffer from leukemia, pseudocrupp or electric smog are not able to participate in normal lifestyles. Modern politics comprises a high level of integration. This is the precondition for internal peace in a society. As turning away from Bacon's principle lets individuals find peace within themselves and a new understanding of autonomy, so does the integration of politics which establishes a new sovereignty.

Thus we can see, if philosophy accepts its responsibility for nature's destruction we can find through philosophy both a new individual selfunderstanding and a new political identity. Without them we do not have a chance for the future.

References

Ackrill J (1974) Aristotle on eudaemonia. In: Proceedings of the British Academy 60, London
Arendt H (1960) Vita activa oder Vom tätigen Leben. Piper, München, Kapitel I
Gloy K (1995) Das Verständnis der Natur. Bd. I: Die Geschichte des wissenschaftlichen Denkens. Beck, München
Gore A (1992) Earth in the balance, ecology and human spirit. Houghton Mifflin, Boston NewYork 1992
Jonas H (1979) Das Prinzip Verantwortung. Suhrkamp, Frankfurt/M.
Kant I (1974) Grundlegung der Metaphysik der Sitten. In: Kant I (1974): Werkausgabe, Bd. VII. Frankfurt/M., p 51
Kane H (1995) Die Heimat verlassen. In: Wordwatch Institute (Hrsg.) Zur Lage der Welt 1995. Fischer, Frankfurt/M., pp 205–236
Maschat M (1989) Leonardo da Vinci und die Technik der Renaissance. Beck, München
Meyer-Abich KM (1997) Praktische Naturphilosophie. Erinnerung an einen vergessenen Traum. Beck, München
Sachs A (1996) Menschenrechte und ökologische Gerechtigkeit. In: Wordwatch Institute (Hrsg.) Zur Lage der Welt 1996. Fischer, Frankfurt/M, pp 196–225
Weizsäcker EUv (1989) Erdpolitik. Ökologische Realpolitik an der Schwelle zum Jahrhundert der Umwelt. Wissenschaftliche Buchgesellschaft, Darmstadt
Welan M (1984) Umweltschutz, Zweck des Staates, Sinn der Freiheit. In: Enrique H. Prat (Hrsg.) Kurswechsel oder Untergang: die ökologische Rettung der Natur. Lang, Frankfurt/M. Berlin, pp 11–14

CHAPTER 3

On Moral Failures in Dealing with Environmental Issues: The Europipe Experience

B.W. Flemming

3.1
Introduction

Environmental ethics may be defined as that part of philosophy that deals with the rights or wrongs (the good or evil) of human interactions with the natural environment. To quote the philosopher Stephen E. Toulmin (1960): "Ethics is everybody's concern". In analogy to this statement it could indeed be claimed that environmental ethics is everybody's concern. It should therefore be viewed as representing a special case of ethics, one that concerns itself with principles of moral behaviour in human attitudes and dealings with nature (cf. Attfield 1983).

In the past, when human pressures on the environment were low, the need to define and implement moral standards in dealing with nature was seldom a conscious issue because human activities rarely resulted in lasting environmental damage. This began to change with the advent of urbanization and political empire building of early civilizations (e.g. severe land degradation associated with the construction and expansion of ancient cities, or widespread deforestation in Roman times). It rapidly expanded with the technological advances made during the late Middle Ages, and took on new dimensions in the wake of the industrial revolution and the associated population explosions in the nineteenth and early twentieth centuries. A further escalation took place after World War II as a consequence of mass production and economic globalization. In spite of this long history of systematic environmental destruction, public awareness of and reaction to this sad state of affairs is a surprisingly recent development.

The memorable book *The Silent Spring* by Rachel Carson (1962) was one of the earliest and most impressive public indictments against uncontrolled exploitation of natural resources and the associated destruction of the environment by the indiscriminent use of pesticides, heralding the forthcoming change in attitude towards nature. From the start, public industrial and governmental reactions were anything but rational. Today we stand in the midst of this (still painful) process which all too often manifests itself in violent confrontation between radical environmental pressure groups on one side and the political and industrial establishment on the other. As a consequence, and contrary to the ideals of nature conservation and environmental ethics, compromises emanating from such confrontations are often counterproductive, both ethics and the environment being the ultimate losers.

In this context the Europipe project is a case in point. In the face of an evidently better alternative (Fig. 3.1), it is today clear that the chosen route through the Wadden Sea and some of the constructional methods applied must in many respects be regarded as an inferior solution from an environmental point of view. Experts generally agree that the most sensible routing of the landfall would have been on the island of

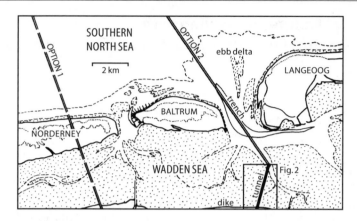

Fig. 3.1. The landfall region of Europipe in the German Wadden Sea. *Option 1* optimal but rejected route across the island of Norderney, *Option 2* ultimately adopted route through the ebb delta and channel system

Norderney, and then across the intertidal watershed between the island and the mainland. This route, intitially the prime choice, would have been the shortest, safest, technically least demanding, and financially least costly of any alternative, including the one finally adopted. In addition, it would have been a less damaging operation from an environmental (and ecological) point of view in spite of claims to the contrary.

How can this seemingly irrational decision be explained? An analysis of the course of events clearly shows that this decision was practically the only remaining option left after years of confrontation between radical and uncompromising environmentalists and the authorities. This confrontation was characterized by wilful disinformation, distortion of facts, gross exaggerations, and scientifically unsupported claims on the one side and political expediency and/or weakness on the other. This can be demonstrated by numerous examples from the start to the end of the project. It illustrates how the degeneration of moral standards in dealing with environmental issues has subverted the valid and honest concerns of the environmental movement (cf. Rowell 1996).

An analysis of the project from the first announcement in 1991 to its final completion in 1995 demonstrates that ethical principles and moral standards were either violated, infringed, neglected, ignored, or simply forgotten in almost every phase and by almost every party directly or indirectly involved. This includes the pipe operator, the politicians, the environmentalists, the fishing community directly affected by the construction, the media, and – last but not least – also the scientific community.

3.2
Moral Failures in the Europipe Conflict

3.2.1
Original Moral Failure of Politics

It may sound surprising that the roots of the socio-political conflict associated with the planning and construction of Europipe can be traced back to the lack of foresight by those politicians who drew up the boundaries of the Wadden Sea National Park and formulated the park regulations. Not that the idea of a national park was misguided. It was the proclamation of the entire Wadden Sea coast (only excluding the shipping lanes of the Ems, Jade, Weser, and Elbe river estuaries) as a national park which was ill advised and shortsighted. By forgetting or being unwilling to reserve a corridor for economic activities outside of shipping lanes (e.g. for the construction of a gas pipeline), the politicians concerned inevitably and predictably laid the foundation for future conflicts.

It is not the purpose of this chapter to discuss the ethics of routing a pipeline through a national park. The fact is that the pipeline was constructed through the park. The position taken here is that such action may be acceptable if the intervention is spatially restricted and of short duration, leaves no tangible trace after completion, represents a local and reversible disturbance of the ecosystem (perturbation), precludes viable alternative routes, and is of substantial national interest. It can be argued that in the case of Europipe all these crucial prerequisites were fulfilled without any serious reservations.

3.2.2
Moral Failure of the Pipeline Operator

The predicament caused by the all-inclusive park boundaries was then compounded by the manner in which the project was initially publicised by the pipeline operator, the Norwegian State Oil Company (Statoil). By failing to promote an image of environmental awareness and to give environmental considerations a high priority from the start, Statoil precipitated immediate and outright rejection by radical environmental groups whose political punch was evidently neglected or grossly underestimated. Although Statoil took great pains in rectifying the situation, the company was not able to restore its credibility in environmental consciousness, at least not for the duration of the project. Thus, because ethical considerations were neglected, a simple tactical mistake resulted in the breakdown of rational discussion based on objective arguments and scientific facts. An ironic twist in this unfortunate turn of events is the subsequent abuse by the media of an inoffensive and well-meant cartoon produced by Statoil itself in which a friendly Troll brings the life-supporting oil pipeline ashore (Fig. 3.2). Trolls are giant spirits of the earth in Nordic mythology who are commonly associated with large-scale natural phenomena (cf. Grappin 1965). As shown in Sect. 3.2.6, this figure was transformed into a polemic anti-Europipe caricature by the local press.

3.2.3
Moral Failure of the Environmentalists

The political decision to route the pipeline through an ebb delta, inlet and channel system of the Wadden Sea, instead of landing it on the barrier island with the shortest tidal watershed (Norderney), was taken after years of massive anti-pipeline campaigns by environmentalists. These campaigns were characterized by the wilful violation of ethical principles, accompanied by a corruption of moral standards along the whole way. Distortion of facts, gross exaggerations, and scientifically unsupported claims, in particular, mark the confrontation. Public (and official) opinion was manipulated by statements which, amongst others, claimed that the Wadden Sea ecosystem would be

Fig. 3.2. Cartoon prepared by Statoil illustrating a friendly Norwegian troll bringing an oil pipeline ashore (trolls are earthbound giant spirits in Nordic mythology who existed before the gods and are associated with large-scale natural phenomena)

destroyed, or that allowing pipe construction would open the door to the industrial exploitation of the Wadden Sea.

However, the fact that the final decision of routing the pipe track through the ebb delta and channel system (rather than across the tidal watershed) was received with little opposition by the environmental movement reveals two other moral failures. First, the environmental activists concerned stand accused of submitting to a political appeasement action, although the solution continued to violate the integrity of the national park. Second, by accepting the final route without effective protest the environmentalists appear to have totally ignored the fact that the subtidal channel ecosystem, about which much less was known than the intertidal ecosystem, would be disturbed at least as severely as by a crossing along the intertidal watershed. The justification for such selective campaigning remains obscure and questions its ethical basis. These examples expose some of the contradictory and sometimes even idiosyncratic attitudes of environmental activists. They illustrate the moral *cul de sac* into which the environmental movement has been manoeuvered over the years because of fundamentalist and uncompromising stands in which there is apparently no room for human needs and aspirations. At the same time they fuel the green backlash (Rowell 1996) without realizing the potentially irreparable damage this may inflict upon both their own cause and the environment.

3.2.4
Second Moral Failure of Politics

The political decision of routing the pipeline through the intertidal flats by tunnelling must be seen as an act of appeasement towards the environmentalists. The construction of a 2.6-km-long tunnel was justified by the argument that it would leave the intertidal ecosystem undisturbed. No such ecological concerns were expressed when it became clear that the planned route required the installation of 48 anchor piles along the channel margins followed by trenching over more than twice that distance through the subtidal channel and ebb delta system. In addition, the available tunnelling method had up to that time never been applied over distances greater than about 1.5 km. The risk and consequences of failure were evidently ignored or played down. Fortunately, the expensive tunnelling procedure (by a factor of 8 more costly than trenching) proved successful. It did not, however, achieve its main purpose, namely to leave the intertidal flats undisturbed. Unforeseen vibrations of the tunnelling machine caused visible compaction of the intertidal sediments above the tunnel track (Fig. 3.3). While

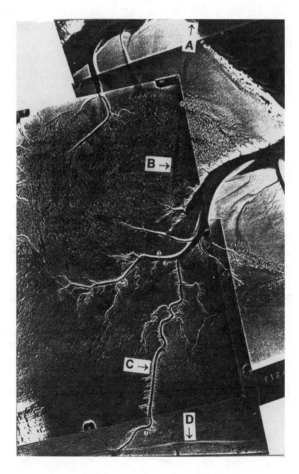

Fig. 3.3. Composite aerial photograph of tunnel section of the Europipe landfall (for location see Fig. 3.1). Clearly visible is the surface imprint of the tunnel track in form of a depression (*B*) and channel (*C*) produced by sediment compaction. *A* indicates location of tie-in chamber and *B* position of landward dike

compaction itself can be a major ecological disturbance factor, it also triggered morpho-dynamic adjustments which culminated in the formation of new drainage networks. Arguments of cost-effectiveness (surely in itself a highly valued principle in economic ethics) are often used to overrule environmental considerations. Ironically, in this particular case the principle was unwittingly turned upside down, i.e., it was neither cost-effective nor environmentally compatible. By capitulating to the political pressure exerted by radical environmentalists, politics must accept a large share of the blame for the escalation of costs associated with the construction of Europipe.

3.2.5
Moral Failure of Local Communities

By uncritically accepting the exaggerations and distortions voiced by the environmentalists, and being driven by their own home-made horror visions, local communities fell victim to their self-generated fears and misjudgements. These included claims that the construction of the pipeline would destroy their home and island, that the region would lose its tourist image and that this, in turn, would result in economic decline, loss of revenue, and increased unemployment. The impact of this misguided public outcry contributed substantially to the political decision of abandoning the original pipeline route (Option 1 in Fig. 3.1).

After construction commenced, it soon became apparant that none of these fears was realistic. On the contrary, the construction actually had the opposite effect on every count, even though the benefits were probably of short duration only. From an economic point of view Europipe certainly did not have any detrimental effects. This did not deter individuals and local authorities from claiming disproportionately high and, in some cases, even totally unjustified financial compensations.

3.2.6
Moral Failure of the Media

As usual, the role of the media was ambivalent. While at national level newspapers mostly reported factually and impartially, the same cannot be said of the local press. The latter produced a mix of factual and emotionalized reporting, mostly with an anti-Europipe slant. Rarely was there a balanced and rational presentation of both sides of the coin, especially in the years leading up to the construction when public concern and the environmental debate were still easily influenced. The polemic caricature in Fig. 3.4 clearly alludes to the Statoil troll (see Fig. 3.2), and is a good example of how the local press fuelled the anti-Europipe debate. In another example, a local author wrote a book narrating a fictitious story about the construction of a pipeline and its detrimental effects on a coastal community (de Witt 1995). The allusion to Europipe is clear, even to the point of adapting the troll to illustrate the book cover.

3.2.7
Moral Failure of Science

As in the case of the media, the position of the scientific community is often ambivalent in environmental conflict situations. Biologists, in particular, appear to be emotionally

Fig. 3.4. Polemic caricature published in the local press to arouse anti-Europipe sentiments in public opinion. Thus, the sympathetic troll illustrated in Fig. 3.2 has been converted into a fat monster bulldozing its way through the Wadden Sea National Park and the district name "Ostfriesland" has been replaced by "Ostgasland"

torn between the rational dictates of scientific fact and their sympathy for the environmental movement. In the Europipe conflict local biologists remained conspicuously quiet when confronted in the media with claims that pipeline construction would destroy the Wadden Sea ecosystem. There is ample evidence to prove that such scenarios are ludicrous. The fact is that the ecosystem was to be severely disturbed in the immediate vicinity of the pipe track, but that the detrimental effects of such an intervention were expected to become obscured within a few years, at the very most. Episodic natural disasters such as severe ice winters or catastrophic algal blooms (Delafontaine and Flemming 1997) have lethal effects that are quite comparable to such localized one-time trenching and dumping of the dredge spoil, but in these cases on a truly Wadden Sea scale. Yet, the Wadden Sea ecosystem has to date always recovered from such natural large-scale catastrophes within a year or two. The reluctance of many scientists to express clear, unambiguous and scientifically well-founded opinions in such circumstances not only damages the image and credibility of science in public perception, but also at the same time does a profound disservice to the environment.

3.3
Discussion and Conclusions

In the case of Europipe the most rational solution ultimately became unviable because moral standards and ethical principles were either knowingly or unwittingly violated in the course of the presentation, discussion, communication, and daily reporting on the issue along the whole way. When Rachel Carson (1962) called for an urgent need to end the practice of "false assurances and sugar coating of unpalatable facts", she was referring to the manner in which the chemical industry manipulated public opinion at the time. Her rational, scientifically well-founded and pragmatic approach to the problem in 1962 ultimately bore positive fruits in that the same industry that originally left nothing untried to discredit her has today integrated environmental concerns into its strategic planning (Lindsay 1997). The pendulum, however, seems to have unfortunately swung to the opposite extreme, as the same deplorable practices of discreditation are now being applied in reverse under the guise of environmental concern and protection. This development is undermining the true and justified cause of the environmental movement. It subverts recent progress made towards environmental awareness in industry, and promotes an anti-environmental backlash which can ultimately only do more damage (e.g. Rowell 1996).

By knowingly submitting to the deceptive arguments and unsupported claims of radical environmentalists, political decision makers must accept a large share of the blame for the implementation of inferior solutions. Although in some cases problems of this kind can be alleviated by the skilful use of conflict psychology in mediating between opposing parties (Donner 1997; Endresen 1997), this approach is inevitably doomed to failure if compromise solutions are rejected because of radical fundamentalist attitudes.

There is an urgent need for the return to rational discourse based on ethical principles and moral standards by everyone involved in or concerned with the issue at hand. Only then will similar confrontations and conflicts, in which the environment is the inevitable loser, be avoided in the future. Environmental ethics must become a matter of the highest priority (see Kinne 1997). Since the existential needs of both man and nature are at stake, the development of integrated and sustainable solutions for the resolution of environmental conflicts is overdue (Brenner 1997; Dahle 1997; Dreiblatt and von Schweinichen 1997; Röchert and Wells 1997). Ultimately, man has no choice but to find ways to co-exist with nature without destroying it, or face doom. As aptly stated by Vollmer and Delafontaine (1997), "...the time for conflict is long passé. Actually, it lasted much too long...".

The most desirable solution would be a globally accepted code of environmental ethics and conduct, coupled with guidelines for environmentally sound practices to be applied and respected by everyone concerned. Constructive suggestions to this effect have been made in a set of recommendations which emanated from the "First International Symposium on Large-Scale Constructions in Coastal Environments" (see Recommendations this Vol.).

References

Attfield R (1983) The ethics of environmental concern. Oxford University Press, Oxford
Brenner A (1996) Ökologie-Ethik. Reclam, Leipzig

Brenner A (1997) Philosophy and man's responsibility for nature. In: Vollmer M, Delafontaine MT (eds) First International Symposium on Large-Scale Constructions in Coastal Environments, Abstract Volume. Forschungszentrum Terramare, Berichte Nr. 1:23–24

Carson R (1962) The silent spring, 25th Anniversary edn, 1992. Penguin Books, Baltimore

Dahle Ø (1997) Technology, sustainability and ethics. In: Vollmer M and Delafontaine MT (eds) First International Symposium on Large-Scale Constructions in Coastal Environments, Abstract Volume. Forschungszentrum Terramare, Berichte Nr. 1:23

Delafontaine MT, Flemming BW (1997) Large-scale sedimentary anoxia and faunal mortality in the German Wadden Sea (southern North Sea) in June 1996: a man-made catastrophe or a natural black tide? German J Hydrogr 49 (in press)

de Witt E (1995) Pipeline. De Utrooper Verlag, Leer

Donner H, Zühlsdorff H (1997) Management of environmental conflicts prior to government intervention – a case study. In: Vollmer M, Delafontaine MT (eds) First International Symposium on Large-Scale Constructions in Coastal Environments, Abstract Volume. Forschungszentrum Terramare, Berichte Nr. 1:16

Dreiblatt D, von Schweinichen C (1997) Sustainable integrated coastal zone management. In: Vollmer M, Delafontaine MT (eds) First International Symposium on Large-Scale Constructions in Coastal Environments, Abstract Volume. Forschungszentrum Terramare, Berichte Nr. 1:33–34

Endresen T-M (1997) Managing environmental conflicts. In: Vollmer M, Delafontaine MT (eds) First International Symposium on Large-Scale Constructions in Coastal Environments, Abstract Volume. Forschungszentrum Terramare, Berichte Nr. 1:17

Grappin P (1965) Germanic lands: the mortal gods. In: Grimal P (ed) Larousse world mythology. Hamlyn, London, pp 357–399

Kinne O (1997) Ethics and eco-ethics. Mar Ecol Progr Ser 153:1–3

Linden E (1997) Poet of the tide pools. Time Magazine 150 (16):102

Röchert R, Wells SM (1997) Coastal and marine protected areas – nature's refuge from large-scale constructions? In: Vollmer M, Delafontaine MT (eds.) First International Symposium on Large-Scale Constructions in Coastal Environments, Abstract Volume. Forschungszentrum Terramare, Berichte Nr. 1:24

Rowell A (1996) Green backlash: global subversion of the environment. Routledge, London

Toulmin ST (1960) An examination of the place of reason in ethics, 2nd edn. Cambridge University Press, Cambridge

Vollmer M, Delafontaine MT (1997) Preface to the Abstract Volume. In: Vollmer M, Delafontaine MT (eds) First International Symposium on Large-Scale Constructions in Coastal Environments, Abstract Volume. Forschungszentrum Terramare, Berichte Nr. 1:7

Coastal and Marine Protected Areas: Nature's Refuge from Large-Scale Constructions

Sue M. Wells · Ralf Röchert

4.1
Introduction

There are now over 1000 marine and coastal protected areas (MCPAs) throughout the world, ranging in size from 1 ha to some 340 000 km² (the Great Barrier Reef Marine Park in Australia). They range from small, highly protected reserves that sustain species and critical habitats to larger multiple-use areas in which conservation and protection of natural processes is coupled with various socio-economic activities and concerns (Kelleher *et al.* 1995).

Despite this progress, in many countries, including major maritime nations in the North, the political will to establish and adequately manage, and the awareness of the ecological and economic values of MCPAs, is still lacking. Less than 1% of the planet's marine surface has been designated as protected, compared to over 6% of the land surface. A study of management effectiveness of a number of MCPAs revealed that about 30% had a 'low' management level, 40% 'moderate' and only 30% 'high' (Kelleher *et al.* 1995). Reasons for poor management include, among others: lack of public support, often because of a failure to get stakeholders involved in the establishment and management process; inadequate enforcement of regulations; and lack of clear organisational responsibilities for management, with confusion between different government agencies. Even where the MCPA concept has been accepted as a central component of marine and coastal resource management, there are numerous obstacles to successful long-term management, including major engineering projects.

This is also true for the Wadden Sea, which is legally designated as protected in all bordering countries. The Wadden Sea is one of the world's most important wetlands and part of the Global 200, the world's most important ecoregions as identified by WWF in 1997 (Olson and Dinerstein 1997). It is one of the few Western European habitats which still resembles its natural state and has an extraordinary importance for wildlife, especially birds. A total of 10–12 million waterbirds of mainly Arctic origin use the 10 000 km² area each year, many of them arriving as a large part of or their whole population. A large proportion of European saltmarsh vegetation, with its unique plant and animal community, is also to be found in the Wadden Sea. Thus, the Wadden Sea acts as a major refuge for nature in industrialised Europe.

4.2
The Case of Europipe

In the Wadden Sea, a large-scale construction, the Europipe, has been located in one of the world's most important wetlands. This area is currently managed under a co-operative governmental arrangement between the countries of Denmark, Germany and The Netherlands. The 'guiding principle' for management, adopted at the Sixth Trilateral Governmental Conference for the Protection of the Wadden Sea in 1991, is "to achieve, as far as possible, a natural and sustainable ecosystem in which natural processes proceed in an undisturbed way". At the same meeting it was agreed that, in principle, construction of new pipelines is to be avoided (CWSS 1992). These principles are policy, however, and have no legal basis.

The German part of the Wadden Sea is designated a 'national park', the strongest conservation category that German environmental law offers. The Wadden Sea is also listed under some of the international and regional treaties which require that contracting parties protect and effectively manage key sites of the world's natural ecosystems. However, neither the legal designation as a National Park nor the tri-nationally adopted principles and international obligations have prevented the construction of the Europipe. WWF's view is that this represents a failure on the part of the German government to enforce national legislation and meet international obligations.

This raises several questions. How 'protected' are protected areas, if detrimental economic interests and political support for them can overturn environmental laws and conservation arguments? Can any large-scale construction in a protected area be considered sustainable use, and compatible with the objectives of such a site? Or conversely, should protected areas not be, at a minimum, free from large-scale industrial development and thus a 'refuge' from major constructions? The Europipe case is not the only problem of that sort, but it provides a starting point for illustrating the national and international obligations that exist for the protection of biodiversity through the establishment and management of protected areas.

4.3
National Obligations

The role of protected areas in protecting the world's biodiversity and contributing to sustainable development is now well established. At the Earth Summit (the United Nations Conference on Environment and Development) in 1992 in Rio de Janeiro, one of the major outputs was Agenda 21, an action plan designed to carry biodiversity protection work through to the next century. Chapter 17 of Agenda 21 covers the oceans and specifically requires that "states should identify marine ecosystems exhibiting high levels of biodiversity and productivity and other critical habitat areas and should provide necessary limitations on use in these areas, through inter alia designation of protected areas". Subsequently, many of the agreed principles in Agenda 21 were specified as obligations for contracting Parties to the Convention on Biological Diversity (De Fontaubert et al. 1996). This requires that, among other things, each Party shall, as far as possible and as appropriate: 1) establish a system of protected areas or areas where special measures need to be taken to conserve biological diversity; and 2) develop, where necessary, guidelines for the selection, establishment and management of such areas.

In order for a protected area to be recognised internationally, it must fall within certain definitions laid out, for example by IUCN ("An area of land and/or sea especially dedicated to the protection and maintenance of biological diversity, and of natural and associated cultural resources, and managed through legal or other effective means") and the Convention on Biological Diversity ("A geographically defined area which is designed or regulated and managed to achieve specific conservation objectives").

The purposes for which protected areas are established vary considerably, which has led to the development by IUCN of an internationally recognised set of categories and criteria by which their management objectives can be defined. For example, marine and coastal protected areas may be established primarily for protecting endangered species or key breeding and migration sites. In many countries, they also play an important role in sustaining commercially or locally important fisheries, reducing conflict between different user groups, regulating tourism and leisure activities, and providing alternative sources of income to displaced fishermen through the creation of jobs in the tourism and recreation industries. In practice, many protected areas are managed for a combination of purposes, often using systems of zoning. Since certain forms of human use are compatible with some of the main functions of protected areas, in 1994, IUCN revised and approved a new set of categories of protected areas (IUCN 1994; Table 4.1).

Categories I–III are mainly concerned with the protection of natural areas where direct human intervention and modification of the environment has not occurred or has been limited; categories IV–VI concern areas with significantly greater intervention and modification. It should be noted that the national names for protected areas vary greatly and may not reflect the international categorisation e.g. National Parks in some countries are assigned to category V, rather than category II, as they contain human settlement and extensive resource use. The Lower Saxonian Wadden Sea National Park will become a category II area, but does not yet qualify and is classified at the moment by IUCN as category V. IUCN emphasises that all categories are needed for conservation and sustainable development, and encourages countries to develop protected area systems using the full range of management categories as appropriate.

Categories IV–VI emphasise the principle that sustainable human use can be compatible with many forms of protected area, and indeed may be highly beneficial for the area. For example, biosphere reserves are in categories V and VI, as their emphasis is on linking the protection of representative natural areas (in the core zones) with sustainable use and ensuring that local people benefit. By broadening the scope of protected areas, they are seen as less threatening as they do not imply a complete block on human activities. However, this can also lead to abuse of the general principles that lie behind the concept of protected areas. The lack of a clear distinction between 'protection' and 'sustainable use' raises the question of whether significant or even adequate protection will be provided for biodiversity.

Nevertheless, there is a generally agreed principle that to meet any of the criteria for designation under one of the IUCN categories, protected areas should exclude industrial-scale activities such as intensive farming and forestry, large-scale mining or large settlements. In many countries, there is legislation to prohibit oil and gas exploitation and exploration in certain categories of protected areas. Where human activities are permitted, these must generally be compatible with the objectives of the protected area itself, such as traditional exploitation activities, or education and tourism. Human activities of any kind are generally strictly regulated in core zones of multiple-use areas, and in protected areas categorised under the most strict protection category.

For example, in the north-east Svalbard Nature Reserve in Norway (designated as category I under the IUCN system), construction, mining and other activities that interfere with the terrain or disturb the natural environment are prohibited (IUCN 1994). In Australia, the threat of oil drilling and sand and limestone mining to the Great Barrier Reef was the trigger for the establishment of the Marine Park. The Royal Commission that reported on the controversy in 1973 concluded that even with the strictest guidelines, there would inevitably be oil leaks that would impact on the reef. In 1975 the Great Barrier Reef Marine Park was gazetted. It is designated as category V under the IUCN system and mining and drilling are prohibited throughout, even in the General Use Zones where a range of other activities are permitted (IUCN 1994). In the UK, consideration is being given to the designation of 'Sacrosanct Areas' which would be closed to all forms of oil, gas and mineral exploration and exploitation on account of their ecological sensitivity and importance.

4.4
International Obligations

Several international and regional agreements now exist which require that contracting parties protect and effectively manage key sites of the world's natural ecosystems. For example, the Wadden Sea is included in the *List of Wetlands of International Importance* under the Convention on Wetlands of International Importance (Ramsar). Under this treaty, contracting parties are required, among other obligations, to promote the conservation of wetlands in their territory through the designation of protected areas (nature reserves). Particular conservation duties pertain to listed sites. Contracting parties are obliged to inform the bureau of any past, present or future ecological changes to listed wetlands, and a monitoring procedure ensures that this is followed and the necessary mitigation actions taken.

The Convention allows for the 'wise use' of wetlands, provided that this does not adversely affect the ecological character of the wetland. Detailed guidelines on wise use have been drawn up which enshrine the precautionary principle. Planning decisions have been influenced in many countries by the listing of sites under Ramsar – with the rerouteing of roads, or the curtailment of development activities. In the case of the Wadden Sea, there have been serious concerns as to whether the Europipe could be considered wise use, and the issue was taken up through the Ramsar mechanism, as reflected in recommendations at the fifth conference of the parties in Japan, 1993, and in other fora. Ramsar sites where ecological changes have occurred, are occurring or are likely to occur may be listed on the *Montreux Record*, if the relevant government gives its consent. The aim of this is to focus attention on the need for remedial action – the site is removed from the record when the problems have been overcome. As yet, the Wadden Sea has not been entered on the *Montreux Record*.

The European Union Directive on the Conservation of Natural Habitats and Species requires states to designate, by the year 2004, sites of international conservation importance as Special Areas of Conservation (SACs), under the Natura 2000 programme. The EU Birds Directive similarly calls for the establishment of Special Protected Areas (SPAs). The Wadden Sea is identified as both an SAC and an SPA under these Directives.

Consideration has also been given to nominating the Wadden Sea as a World Heritage Site under the UNESCO World Heritage Convention. Under this Convention, natural and cultural sites of 'outstanding universal value' are designated, provided their management system meets strict protection criteria. A monitoring programme on the conservation status of the listed sites allows threats to be identified which, if serious, can lead to the site being inscribed on the *List of World Heritage in Danger*, a process similar to the *Montreux Record*. A nomination for the Wadden Sea has yet to be prepared.

Three areas of the Wadden Sea, including the Lower Saxonian Wadden Sea National Park, are designated as biosphere reserves. The Man and the Biosphere Programme is not a convention, but an international programme aimed at conserving representative natural areas. Criteria for designation as a biosphere reserve include representativeness, diversity, naturalness and effectiveness as a conservation unit. Biosphere reserves are generally managed with a highly protected core area and a surrounding buffer zone.

4.5
The Ethical Argument: Respect for Nature

Equally importantly, but promoted globally to a far less extent, is the ethical argument for maintaining protected areas as 'refuges' from human activities on a large-scale. The 1991 revision of the IUCN/WWF/UNEP world conservation strategy, *Caring for the Earth*, took a major step forward in recognising openly that if human behaviour is to change sufficiently to preserve the planet's life support systems and a significant amount of its biodiversity, a universally shared set of ethical values is needed (IUCN/UNEP/ WWF 1991). A 'world ethic for living sustainably' was therefore developed, using principles of natural science, democratic social traditions, and ethical and religious traditions:

> "All life on earth, with soil, water and air, constitutes a great, interdependent system – the biosphere. Disturbing one component can affect the whole. Our survival depends on the use of other species, but it is a matter of ethics, as well as practicality, that we ensure their survival and safeguard their habitats.

> ▪ Every human being is part of the community of life, made up of all living creatures.
> ▪ Every human being has the same fundamental and equal rights.
> ▪ Each person and each society is entitled to respect of these rights.
> ▪ Every life form warrants respect independently of its worth to people.
> ▪ Everyone should take responsibility for his or her impacts on nature.
> ▪ Everyone should aim to share fairly the benefits and costs of resource use.
> ▪ The protection of human rights and those of the rest of nature is a worldwide responsibility that transcends all cultural, ideological and geographical boundaries."

This draws on basic tenets of the UN Universal Declaration of Human Rights, and the UN World Charter for Nature, which recognise that "every life form is unique, warranting respect regardless of its worth to humankind". Thus, the world ethic is not a radical new approach. The concepts of 'respect for nature' and the recognition that all life forms have intrinsic value, are deeply rooted. The scientific perspective also emphasises the ethical significance of the holistic, community or system-based approach, i.e. that all species are components of our life support system. Promotion of such an ethic is viewed as essential, since people's behaviour is dependent on their beliefs, and widely shared beliefs are often more powerful than government legislation.

At the same time, it has to be recognised that there is inevitable conflict between human rights and the rights of nature, between meeting human needs or demands and preserving biodiversity. Typically, there are few concerns by industry for non-binding principles, as in the case of the trilateral Wadden Sea management framework, if economic gains or political pressures are sufficiently high that they can be ignored. However, under the world ethic for living sustainably, industry itself should aim to avoid using protected areas for construction work, should seek technical alternatives, and should not take advantage of situations where legislation is weak or enforcement is poor. In the case of pipelines, alternative routes should be a serious option, *even* if more expensive. Furthermore, the ethic recognises that with rights go responsibilities, and that everyone (governments, industry and individuals) must take responsibility for resolving such conflicts, and helping to maintain protected areas for the purposes for which they were designed. A greater 'respect for nature' would mean the recognition that certain areas should remain unviolated on principle, for their beauty, uniqueness and ecological integrity.

4.6
Conclusion

Current trends in conservation and management of the natural environment have been to emphasise the role of sustainable human use and development. This is essential in many cases, where human survival and quality of life is dependent on natural resources. However, there are strong arguments for ensuring that nature conservation takes precedence over economic interests in protected areas. Industry should respect these and, on ethical grounds alone, avoid using protected areas for large-scale constructions. WWF's view is that both governments and industry must take greater responsibility for enforcing protected area systems and maintaining them in an unviolated state. Once one form of incursion such as Europipe has occurred and been sanctioned, subsequent 'abuses' are much easier, and the original objectives of the protected area will be progressively eroded. There are few human activities which in the long-term carry no risk, and the application of the precautionary principle, a central tenet for sustainable resource management, is a high priority in the management of protected areas.

In addition to the ethical argument that some natural areas should be set aside as pristine areas to maintain their ecological integrity for future generations, there are also scientific arguments. Firstly, large-scale constructions by their very nature may ultimately have a negative impact on that environment, and, secondly, there is a need for truly pristine areas as reference sites for the future. Governments must improve enforcement of their protected area systems and meet their international obligations in this regard. There must be greater political commitment to ethical principles and a willingness to support these beliefs with appropriate legislation.

As pristine coastal habitat and unpolluted and unfished marine waters become increasingly restricted in size, and the competing interests of industry and development intrude even more, the question of how 'protected' the world's protected areas are will become increasingly urgent.

Acknowledgements. We are very grateful to Hans-Ulrich Rösner and to other colleagues at WWF-Germany and in the WWF Marine Advisory Group for their assistance with this chapter.

References

CWSS (1992) Sixth Trilateral Governmental Wadden Sea Conference – Ministerial Declaration. Common Wadden Sea Secretariat (CWSS), Wilhelmshaven

de Fontaubert AC, Downes DR, Agardy TS (1996) Biodiversity in the seas: implementing the Convention on Biological Diversity in Marine and Coastal Habitats. IUCN, Gland and Cambridge

IUCN/UNEP/WWF (1991) Caring for the earth: a strategy for sustainable living. IUCN, Gland, Switzerland

IUCN (1994) Guidelines for protected area management categories. CNPPA, with WCMC. IUCN, Gland, Switzerland and Cambridge, UK. 261 pp

Kelleher G, Bleakley C, Wells S (eds) (1995) A global representative system of marine protected areas. The Great Barrier Reef Marine Park Authority / The World Bank / The World Conservation Union (IUCN). 4 vols

Olson DM, Dinerstein E (in press) The Global 200: a representation approach to conserving the earth's distinctive ecoregions. Conservation Biology

Part II
Environmental Policy

Towards Developing Guidelines for Assessment, Design and Operation of Estuarine Barrages

T. Neville Burt · Ian C. Cruickshank · Mervyn A. Littlewood

5.1
Introduction

Many large-scale barrage schemes are under consideration both in Europe and world-wide. Some have been built, some are under construction and some are in their design development stage. The authors have been involved in feasibility studies for five of them from London's Thames Barrier in the early 1970s to the Cardiff Barrage in more recent years. Whether we call these structures barriers, barrages or, in some cases, weirs, they are all, for various reasons, designed to modify or totally prevent the progression of the tide up an estuary or inlet. This change to the natural tidal progression has wide ranging implications on the estuary regime. Therefore setting the environmental policy for barrages epitomises the problems associated with any large-scale construction in tidal estuaries. At present there is little in the way of design guidance to help the developer or the regulators. It seemed an appropriate time to review what had been learnt in order to provide environmental policy guidelines for future schemes and identify research priorities to achieve sustainable development.

A research project was set up at HR Wallingford to achieve this, funded by the UK Department of the Environment and the Environment Agency. Information has been gathered from as many barrage projects as possible worldwide and is currently being assessed by a number of specialists. The end product will be published guidelines on the assessment, design and operation of estuarine barrages covering a wide range of issues under the broad headings:

- Barrage concepts, planning and legislation
- Fisheries
- Conservation
- Hydrodynamics
- Morphology
- Flooding
- Groundwater
- Water quality
- Navigation
- Waves
- Structural engineering concepts
- Water-related risk

Environmental science is still developing and we do not yet fully understand all of the processes involved. With the advent of the Rio Summit and conservation directives such as the Habitats Directive it is more important than ever to try to understand these processes and thereby implement best practice. Indeed, future barrage schemes need to closely embrace the principles of sustainable development, thus ensuring that economic investment and environmental improvement go hand in hand. In assessing the hydraulic feasibility and environmental impacts of these schemes, some of the issues required the development of new study techniques to provide predictions of the effects; some issues were virtually unresolvable with present knowledge and even when established study techniques were used there is still the question of how much confidence could be placed in the results. It is hoped that the guidelines will help facilitate an integrated approach to the planning and implementation in which the interests of all the stakeholders are considered and carefully balanced.

This chapter gives an overview of the research project and some of the key issues for various types of barrages, drawing on illustrations from the authors' research. It makes a few brief recommendations for future barrages in lieu of these guidelines becoming formally available.

5.2
Definitions and Purposes of Barrages and Barriers

In order to discuss the environmental issues of barrages it is first important to define exactly what a barrage is. The definition is open to some debate. The generally accepted definition of a barrier and a barrage is as follows:

A *barrage* is an obstruction which permanently excludes all or part of the tidal flow from all or part of the estuary and through which the freshwater flow down the river can be passed by sluices or weirs when the tide is lower than the level of the impounded water behind the barrage. A *barrier* (sometimes refered to as a tidal surge barrage) can be considered as a device which is operated only to prevent exceptionally high tides and surges from passing up the estuary when a surge flood warning is issued: it may be a permanent structure with moveable gates or it may be of fabric design.

The boundary between the two definitions is not so clear when considering some of the present day multi-functional structures which are capable of achieving all permutations of tidal and fluvial flow control. Barrages can have significantly more impacts on the environment than barriers because barriers only temporarily change the natural regime. It is therefore important to retain the distinction between these two types of structure and the guidelines therefore focus on barrages whilst noting particular issues that relate to barriers. The purposes of the construction of barrages or barriers can be described as aiming to achieve one or more of the following objectives, each of which may be applicable in varying degrees:

1. *Improved amenity value* barrages are constructed to improve the amenity value of the impounded water and the surrounding land. This may be in the form of: increased water area for recreational use; improved access to the water area; more stable water conditions providing opportunity for an aesthetically pleasing waterside development where both business and housing can thrive. These barrages are often associated with urban regeneration projects (especially in the UK) where it is seen that the improved amenity value will act as a catalyst for regeneration. The economic aim is for the barrage to increase land values which will offset the capital cost of its construction.

2. *Tidal surge protection* barrages (barriers) are constructed to reduce the risk of flooding from tidal surges. They may be permanent or only operated occasionally. Where they are operated occasionally they would not be expected to have any significant long-term effect upon tidal propagation or the overall hydraulic regime of the estuary. However, the highest tides will never be allowed through so the absolute tidal limit moves downstream, with possible effects on the tidal limit ecology. It is also probable that there will be a temporary local impact which could change the current velocity or bed sediment distribution. Within their design life it is also probable that the barrage will be used with increasing frequency to combat the predicted rise in mean sea level.

3. *Tidal power generation* barrages are constructed to provide electricity generation from the tidal movements and are usually constructed in areas of high tidal range (>4 m). This type of structure deliberately modifies tidal range and propagation to extract energy.

4. *Water storage* barrages are constructed to provide a freshwater lake reserve for abstraction. All tides are totally excluded. Some are to provide a minimum depth for cooling water abstraction (e.g. original Clyde Weir).

5. *Improved water quality* barrages aim to improve the flushing characteristics within a multi-outlet estuary to improve water quality, for example in the case of the Po River Delta in the northern Adriatic.

6. *Silt exclusion.* A particular problem sometimes occurs in the upper tidal reaches of estuaries with high concentrations of silt. During periods of low fresh water flow, silt migrates upstream and deposits, causing a reduction in the cross-sectional area of the estuary and its capacity to deal with a fluvial flood.

7. *Traffic management.* The City of Hull has a particular problem in that the River Hull is used for commercial vessels but is a barrier to road traffic. The main road bridges are low level and such is the nature of the tides that vessels requiring maximum depth have to navigate the river at the same time of day as peak road traffic occurs. This requires the bridge to be lifted causing severe delays to commercial traffic. One option being considered is a barrage to maintain a navigable depth at all times, allowing flexibility of traffic management.

8. *Multi-function barrages.* Most barrages aim to achieve more than just one of the above objectives. However, the majority of barrages have their primary aim as satisfying only one of the above objectives. Where the primary aim of the barrage construction consists of more than one of these objectives then the barrage may be described as a "multi-function barrage". Each promoter will have different objectives for a barrage. The main promoters of barrages in England and Wales are:
   - Development corporations, e.g. Cardiff Bay Development Corporation (Cardiff Bay Barrage)
   - Local government, e.g. Greater London Council (Thames Barrier), Swansea City Council (Tawe Barrage), Kingston upon Hull City Council (Hull Barrage)
   - Private developers, e.g. Hayle Harbour barrage developers (Hayle Harbour Barrage)
   - Power generation companies and consortiums, e.g. Nova Scotia Power Corporation (Annapolis Royal Tidal Power Station) and The Severn Tidal Power Group (Severn Barrage)

A barrage will become a permanent feature, so the long-term ownership and management issues are as important as the construction itself. This is especially true in the case of development corporations which only have a finite life. The ownership will vary depending on the promoter and the particular circumstances surrounding each development. The construction of a barrage will affect many people's lives, and it is important to consult widely. In England and Wales the interested parties/stakeholders will include:

- The Environment Agency
- Ministry of Agriculture, Fisheries and Food
- Local authorities
- Port authorities
- English Nature / The Countryside Council for Wales
- The Royal Society for the Protection of Birds
- The Royal Yachting Association (RYA)
- Utilities (water companies, etc.)
- Local watersports clubs
- Private companies
- Sea fisheries committees
- Fishery owners and angling associations
- Crown Commissioners
- Navigation authorities
- Private individuals
- Landowners
- Water user groups
- The Countryside Commission

Interested parties/stakeholders and the public should be involved early in the planning stage and be encouraged to seek consensus on the acceptable impacts and appropriate compensation measures. They will participate in the planning procedure in a variety of ways. Many will be consulted in the planning process, but should their interests not be satisfied at this stage then there are formal routes available to object to the scheme. These routes are discussed in detail in the guidelines.

Perhaps the main 'interested party' with respect to barrage developments in England and Wales is the Environment Agency. The Environment Agency acquired the duties and responsibilities of the National Rivers Authority (NRA), the Waste Regulation Authorities, and Her Majesty's Inspectorate of Pollution for England and Wales under the Environment Act 1995. The statutory guidance (November 1996) on the Environment Agency and sustainable development lays down its role. The statutory duties are undertaken both by 'influencing' the decision makers and by regulating the environment through the issuing of consents, licences and approvals.

5.3
The Research Project

The UK Department of the Environment (DoE) commissioned HR Wallingford (HR) to "learn from experience the essential elements of barrage design and consolidate the knowledge into guidelines for future application. The target audience is local authorities, the National Rivers Authority (now the Environment Agency), development corporations and any others who may wish to promote a barrage scheme and consulting and contracting engineers who will be responsible for the design and construction". The aspects to be considered are listed in the contract and are reflected in the chapter headings of the guidelines. The general methodology was to:

- Review previous HR studies and data
- Hold discussions with those responsible for the operation of existing structures and those affected by them
- Assess monitoring data where it is available
- Collate and report findings
- Produce guidelines (currently being finalised)

The deliverables were:

- A report on the studies carried out
- Published guidelines (aiming to be published next year)
- A seminar/workshop
- Conference – the First International Conference on Barrages, 11–14 September 1996 (Burt and Watts 1996)

The other funding was provided by the Environment. The overall objective of the Environment Agency contract was "to review available experience on the design, operation and environmental impact of estuarine barrages in order to provide best practice design and operational engineering solutions to overcome or mitigate problems and enhance, where possible, the aquatic and riparian environment". The specific objectives were:

- To review available experience in the UK and abroad on the design, scheme development, operation and environmental impact of estuarine barrages
- To give consideration to barrage design and operation in relation to the aspects listed (broadly similar to the above DoE list) and, in particula, to the following environmental issues: siltation, passage of migratory fish, stratification, temperature, modelling, barrage ecosystems
- To identify current best practice design and operational engineering solutions to overcome or mitigate problems and to enhance, where possible, the aquatic and riparian environment
- To identify those areas where future research is required to improve and supplement existing engineering practice and the understanding of the environmental impact of barrages
- To produce guidelines to assist barrage developers, planning and environmental protection agencies in the design and management of existing and future barrages. This will help ensure an integrated approach is adopted in the planning and implementation process

Careful consideration was given to the format of the guidelines. The decision, agreed with the Environment Agency, was that the guidelines should be issue-based rather than trying to present the development of a barrage project in chronological order. The main generic headings formed the chapter headings, as outlined in the next section.

5.4
The Issues

It is not possible within the constraints of this chapter to fully discuss all the issues related to barrages. It is, however appropriate to provide an overview of the complex interaction of the issues under their generic headings.

5.4.1
Fisheries, Nature and Conservation

Clearly, a barrage whether it fully or partially excludes the tides represents a significant change to an estuary. The most obvious initial impacts are on the hydrodynamics as already discussed, but estuaries have developed biotas, perhaps over centuries suited to the peculiar characteristics of the hydrodynamics as well as climatic and geophysical conditions of the area. It follows that if the hydrodynamics change, the biota will also have to adapt, die or vacate the existing habitat and allow new species to move in. In addition to this, the construction of a barrage will act as a physical obstruction to the passage of migratory fish.

Biologically, estuaries rank amongst the most productive of natural systems. Primary producers include phytoplankton, free-floating in the water column, microalgae attached to the surface of sediment and larger plants and macrophytic vegetation such as saltmarsh and eel grass. Macroalgae (i.e. the typical seaweeds of rocky, coastal shores) are usually sparsely distributed in estuaries due to the lack of availability of suitable hard substrates for attachment.

Fig. 5.1. Fish pass on Tawe Barrage, Wales, showing the highly turbulent flows which significantly affect provision for free passage of fish. (Courtesy of HR Wallingford)

Although a substantial number of predictive studies have been undertaken with regard to the likely effects of barrages, relatively little information has yet been forthcoming on the actual impact of implemented schemes; a situation which somewhat constrains debate on the likely effects of proposed developments. However, a change from a brackish estuary to a freshwater lake will have many obvious changes on the ecology! The present account draws primarily on predictive studies and is, therefore, largely theoretical, but some information on the actual impact of the Tawe (partial exclusion) and Tees (total exclusion) Barrages is presented in the guidelines (see Fig. 5.1).

The environmental policy of zero change in ecology is an impossible target and efforts must be directed towards identifying the key issues and deciding what degree of change is acceptable or beneficial and indeed sustainable. This will result in some mitigation measures and hopefully some enhancements in the development of a scheme. The potential impact of barrage developments in tidal environments on fisheries and nature conservation interests is of major concern to the Environment Agency because of the Agency's fundamental responsibilities in both of these areas. It is also a concern to anglers, fishery owners, conservation groups and the general public.

It is the responsibility of the barrage scheme promoters to undertake an environmental assessment of their scheme, should this be required, and to present the results of their studies as an environmental statement. This document should be prepared in a manner suitable to answer any questions posed by the Environment Agency. For their part the Environment Agency must be aware of the likely impacts of barrage developments on fisheries and nature conservation and have a view on what constitutes reasonable good practice for the assessment of such impacts, i.e. defending the rights of legitimate "users" of the environment. This latter point is of particular importance as reliable methods to precisely predict the impact of barrage schemes on fisheries and nature conservation interests do not exist.

5.4.2
Hydrodynamics

The construction of a barrage can alter the tidal hydraulic regime in a number of ways. It is important to considers the effect a barrage is likely to have on the main hydrodynamic processes that control levels, flows, mixing, flushing and transport of dissolved and suspended matter in the separated water bodies upstream and downstream of the new structure. The issues relevant to hydrodynamics can be summarised as follows:

- The truncation of an estuary by a barrage can radically alter the tidal regime downstream specially if the estuary is close to resonance.
- The pattern of saline intrusion, stratification and gravitational circulation, pollution, transport and deposition of fine sediments in a deep estuary can be sensitive to the changes in tidal flows caused by the construction of a partial or complete barrage and by the extraction or rerouteing of non-saline inflows.
- A poorly located barrage in a shallow estuary (where tidal range approximately equals depth) can cause major siltation problems downstream.
- Half tide barrages which are designed to hold up low-water levels to prevent the exposure of unsightly mud banks and barrages with navigation locks, which allow the intrusion of seawater, give rise to a whole class of issues in the upstream pool.

- The intruding seawater forms a stable two layer system with little or no vertical turbulent exchange. Mud and polluted particulate matter tend to be trapped in the lower layer causing major water quality problems. There can also be significant chemical exchange across the sediment/water interface.
- It is difficult and expensive to design and operate gate control and selective withdrawal strategies which are effective at flushing the landward extremities of the lower polluted layer without nullifying the benefits of the barrage.
- Construction works can have significant adverse effects in terms of erosion and deposition of marine muds.
- A barrage may damage the fauna and flora, by whatever cause, both upstream and downstream of a barrage and this is a main issue to be considered during construction and operation. There is particular concern for the effect of reduced tidal flows and regulation of freshwater inflows on mixing and flushing of dissolved and suspended pollutants. Another area of concern is the potential changes in the patterns of erosion and deposition of fine sediments. Changes in the hydraulic regime will also impact on the behaviour of migratory fish.

5.4.3
Morphology

The morphology, or shape, of an estuary depends upon the fluvial and tidal flow within the estuary and the fluvially and tidally borne sediments. In a natural estuary, the morphology results from a balance between the flow and sediment movement. The river brings fluvial sediment into the estuary while tidal flows introduce tidally borne sediments from the sea. The construction of a barrage disturbs this natural equilibrium. The action of a barrage is to affect the flow that is responsible for the movement of the sediment and also to directly affect the movement of sediment at the barrage site itself. The impact on the flow may be local to the barrage, leading possibly to local scour, or it may be general, as, for example, when the barrage maintains a permanently high water level upstream. In this latter case the impact of the barrage on the morphology of the river may extend for some distance upstream. By modifying the flow in an estuary the impact on the morphology may also extend a significant distance downstream.

The barrage usually increases water levels upstream. By excluding tidal flows, a further impact of a barrage is usually to reduce the discharge at the barrage itself. These effects reduce the hydraulic gradient upstream and hence reduce the sediment-transporting capacity of the channel upstream of the barrage. The sediment production from the catchment remains the same as in the pre-barrage conditions and so the balance between sediment production and transport is disturbed. The reduced transport capacity upstream leads to sediment deposition upstream of the barrage. This can adversely affect flood levels and can affect other users of the estuary. Any reduction in flows at the barrage itself may lead to sedimentation in the channel downstream. This can have an impact on other activities in the estuary and adversely affect flood levels.

By retaining water upstream, a barrage normally reduces tidal discharges downstream. This disturbs the balance between channel size and shape and the discharge and so a barrage induces morphological changes downstream. A barrage may also act as a physical barrier to the movement of coarse sediment, reducing the movement of fluvial sediment downstream. This can affect the nature of the bed sediments both upstream and downstream of a barrage, which can itself result in changes to the morphology.

5.4.4
Flooding

A barrage will affect the flood defence regime in the watercourse on which it is located either as its primary function as in the case of the Thames Barrier or as a consequence of a desired change in water regime such as the Cardiff Bay Barrage (see Fig. 5.2). The principal issues involved in flood defence are:

- Adequate capacity of the gates to pass the design flows with acceptable head loss
- Upstream storage capacity whilst gates are closed
- Adequate and documented training and hand over from the designer and contractor to the barrage operators

Fig. 5.2 One of the physical models at HR Wallingford used in assessment of flood risk on the Cardiff Bay Barrage, Wales. (Courtesy of HR Wallingford)

- Change in flood regime upstream (tidal to fluvial dominance, peak flow to flood volume as critical factor)
- Flood forecasting and control of the structure
- Groundwater seepage from higher impounded levels
- Downstream reflected wave – and operational procedures to minimise this
- Changes to the discharge control of surface water drainage systems
- Changes in the river bed level resulting from the deposition of sediments
- Anticipated rises in sea level

Flood defence is one of the main functions of river authorities including the Environment Agency, which is responsible for the main rivers in England and Wales. The Agency is likely to "require developers to undertake mitigation works at their own cost as a pre-condition to agreement and seek to ensure that development does not proceed until such works are implemented". At all stages of a barrage project from the initial pre-feasibility studies through to the construction and operation, flood defence is one of the key issues that has to be considered.

The first issue to consider is that the regime will change as the relative influence of tide and river (fluvial) flows change. Upriver of a tide-excluding barrage, especially one built near the mouth of a river, the natural river cross-section is generally large because of its previous need to accommodate both tidal and river discharge. The channel itself is therefore likely to have more than adequate capacity to pass the design fluvial flood. The flood risk is therefore almost entirely dependent on the ability of the structure itself to pass the design flood. It is therefore essential that adequate discharge capacity is provided within the structure. Of course fluvial flows can only be passed when the tide level is below the impounded level (i.e. when the barrage is not "tide locked") and the available head for discharge may be limited by the tide level.

It may be possible to drain down the water ahead of any anticipated flood to maximise the storage capacity and minimise the risk of flooding. In many cases it may also be considered undesirable to drain down an impounded river/estuary because of the effect on structures, moored vessels, etc. If a drain down procedure is to be used a good warning system will have to be adopted. Closing a tidal surge barrier at low water can be of great advantage to reduce the tidal restriction on the evaluation of fluvial floodwaters if the structure is far enough downriver to provide a sufficient storage area to accommodate the fluvial flood discharge until the barrier can be re-opened on the following falling ebb tide. The Hull tidal surge barrier, for example, has been used in this way, closing the barrier at low water, thereby providing the whole tidal volume as storage for the flood volume. Drains upriver of a proposed barrage will, in many cases, be controlled by tide flaps, impounding drainage water during tide-locked periods and discharging at low tide. If a permanent high water level is created by a barrage, this method will cease to work and pumping or diversion will be necessary.

5.4.5
Groundwater

Barrages affect the water levels in the estuaries they cross. Historically, levels in surface water channels have been studied by hydrologists and sub-surface flows by hydrogeologists without any real concern for the exchange between surface water and sub-surface waters. However, these exchanges do occur and their importance under certain conditions, such as those which could result from the construction of a barrage, may not be negligible.

Tidal barriers and barrages affect the water level in an estuary and since there is interaction with the aquifer they can affect the groundwater levels. Tidal surge barriers have little impact on the mean water levels in the estuary and hence little impact on groundwater. Power-generating tidal barrages will probably have some, if limited, impact, but water storage tidal barrages and multi-functional tidal barrages will result in an increase in mean water level in the estuary, primarily landward of the impounding structure, and hence an increase in groundwater level. The increase in groundwater level could be widespread, but, except for immediately adjacent to the estuary, it could take a considerable time to develop. The rises in mean water level in the estuary, landward of the impounding barrage structure, are likely to result in conditions which mean there will always be a flow of groundwater away from the estuary.

As well as affecting groundwater flow, water storage tidal barrages and multi-functional tidal barrages can result in changes in groundwater quality. The shallower water table will result in surface pollutants reaching the groundwater in a shorter time. Flow from the estuary can result in groundwater pollution from the surface water and also the spread of pollution away from the estuary from pollution sources such as waste disposal sites and contaminated land. Saline intrusion into aquifers may occur if high levels of saline water are retained upstream of barrages. The changes in salinity will result in a small change in density of the groundwater. This change could affect the load-bearing capacity of the soils, but any affect is likely to be significantly smaller than that which would occur as a result of increases to groundwater level due to the barrage.

5.4.6
Water Quality

Both upstream and downstream water quality may be affected by the introduction of a tidal control structure. Tidal flushing will be altered and salinity variations will change. The scale of change will depend on the location, design and mode of operation of the barrage or barrier. A flood protection barrier which is operated infrequently for a relatively short period of time is unlikely to have a significant impact on water quality. A fully tide-excluding barrage will have an obvious impact upstream through exclusion of seawater and the cessation of tidal flushing. A tidal overtopping barrage can result in saline stratification which could have the most serious impact on dissolved oxygen.

Significant reductions in tidal flushing upstream of the barrage will result in longer residence times and will tend to increase the risk of algal blooms (see Fig. 5.3). Any deterioration in water quality in the impoundment, from whatever cause, is likely to impact on the behaviour of migratory fish. Changes in salinity may impact on the overall ecosystem. A change from saline or brackish water to a freshwater environment will have obvious impacts on the ecosystem. However, sudden changes in salinity caused by intermittent ingress of saltwater into an otherwise freshwater environment may have a serious consequence and prevent the development of a stable ecosystem.

Any significant reduction in the volume of water moving upstream of the proposed site (i.e. the tidal volume) will result in lower tidal velocities downstream. In deep estu-

Fig. 5.3 Algae blooms which have developed since construction of the Wandsbeck Barrage, on the north east coast of the UK. (Courtesy of HR Wallingford)

aries there may also be a change to gravitational circulation which may either compensate for the reduction in tidal flushing or further reduce flushing. Any significant reduction in flushing will tend to make the system more sensitive to pollutant loadings.

5.4.7
Waves

In many barrage situations waves will not present a problem and probably the only provision necessary will be protection against bank erosion. It should not be overlooked that some barrages, by impounding large tracts of water at high water level, create a fetch which could give rise to significant wave effects when the wind direction coincides with the old river alignment. In more exposed situations waves can be a significant factor. For example, in the case of the "Delta" surge barrier on the Dutch coast the structure is exposed to the direct attack of waves generated in the North Sea which has had to be taken into account in the design. At Cardiff it has been necessary to provide a harbour refuge to shelter small boats during storms, whilst they are waiting to enter the bay through the barrage locks. During construction wave erosion of temporary bunds was an issue.

5.4.8
Navigation

A fully tidal excluding barrage will prevent movement of vessels unless some form of lock or removable gate is provided. A part tide-excluding barrage or tidal surge bar-

Fig. 5.4 Navigation lock on the Tees Barrage, UK. (Courtesy of HR Wallingford)

rage may allow shallower draught vessels to pass over it at high water. All three types of barrage will modify navigation to different degrees depending on the specification, situation and barrage design. If commercial navigation is an important issue it will probably be necessary to provide reasonable unrestricted access during all phases of construction as well as in the final scheme (see Fig. 5.4).

5.4.9
Structural Engineering Concepts

In many ways the purpose of the engineering works is to put into practice the requirements made by the environmental policy makers. These include providing:

- An adequate flood discharge system, spillways, training walls, gates and valves and stilling basins
- Locks or alternative navigation routes through the barrage
- Lead-in jetties and dolphins including lead-in structures and dolphins to ensure smooth travel through the barrage and that the fabric of the barrage is protected from vessel impact
- Fish passes to ensure that the passage of migratory fish is not hindered by the barrage
- Beds and bank protection to protect the surrounding areas from a changed regime
- Land reclamation for economic development
- Embankments and walls that actually form the barrage
- Closure of the waterway during barrage construction

Fig. 5.5 Large-scale temporary works required for construction of locks and sluices on the Cardiff Bay Barrage, Wales. Note ship passing the works *(top)* highlighting the fact that tidal water level is several metres above founding level for the works. (Courtesy of HR Wallingford)

- Leisure facilities e.g. canoe slalom
- Temporary works for the above elements (see Fig. 5.5)
- Operating system that takes into account the economic and environmental policy of the barrage (e.g. whether water in times of low flows should be directed to flow down the fish pass or the canoe slalom)

Achieving all these objectives is quite a challenge in itself. Such challenges include enabling vessels free travel through the barrage whilst preventing the flow of saltwater that the vessels are floating in!

5.4.10
Water-Related Risk

Risk assessment has, in recent years, become more widespread as an appraisal methodology. In risk assessment, all the factors producing an event or undesirable outcome are considered for a particular activity or project. The risk may be judged as acceptable or unacceptable according to how frequently an event happens and how much "damage" the event does. Risk assessment provides a framework within which the potential benefits and hazards of a project or course of action can be determined using either a quantitative or a qualitative approach.

There are many different aspects of risk which should be considered at various stages in the barrier/barrage design process, including environmental, economic, structural and flood risks. General information and definitions concerning risk analysis and management are given in the guidelines. However, in brief, the guidelines consider in detail water-related risks including: hydrometeorlogical risk, construction risk and operational risk.

5.5
Environmental Policy

Environmental policy must take account of the issues outlined in the previous section. This section examines the environmental policy framework under which the barrages must be planned, constructed, operated, modified and maintained. Although much of these are UK-based, they are usually derived from aims of the EU directives and/or the Rio Summit striving for sustainability and biodiversity.

5.5.1
Environmental Assessment and Environmental Statement

Environmental Assessment was formally introduced into the UK by the Town and Country Planning (Assessment of Environmental Effects) Regulations 1988 (SI 1988 No. 1199) which implemented much of the European Community Directive on the assessment of the environmental effects of certain public and private projects on the environment (85/337/EEC). Environmental Assessment aims to ensure that any significant effects of new development on the environment are assessed before they take place. The regulations classify barrages as Schedule 2 developments which may require a formal Environmental Statement (ES) if they are to have "significant" effects.

Prior to a full Environmental Assessment, the Environment Agency would expect to see a scoping document based on "Further Guidance on the Environmental Assessment of Project" to ensure that an Environmental Assessment will cover all of the issues concerned. The Environment Agency would expect to see an Environmental Assessment undertaken for all barrage schemes. The assessment would address the effect that the barrage construction will have on the environment. In order to fulfil its basic aim, environmental assessment should be undertaken in parallel with project design. It will normally examine the following issues under the following broad headings:

- Hydrology
- Morphology
- Flooding
- Groundwater
- Water quality
- Fisheries

- Conservation
- Noise
- Dust
- Navigation
- Water resources
- Recreation
- Landscape and archaeology

The Environmental Assessment would include:

- Description of the barrage development
- Identification and evaluation of the main environmental effects
- Description of the affected environmental features and habitats as listed above
- Measures to avoid, reduce or remedy effects

In addition, assumptions and predictions used in environmental assessments should be periodically reviewed for the purpose of taking immediate remedial action.

5.5.2
Environment Agency and Licences, Consents and Approvals

The Environment Agency has statutory duties as the primary environmental agency operating in England and Wales. In relation to barrage developments, the legislative framework within which the Environment Agency operates is primarily that of the UK Water Resources Act 1991 and of equal importance the Environment Act 1995. In relation to barrage construction, a licence/consent/approval will normally be required for:

- Impoundment of water
- Works on the bed or banks of a river or any other construction that is likely to impede flow
- Changes to navigation
- Interruption to the free migration of fish
- Water abstractions
- Trade effluent

The loss of free access for migratory fish is of concern to the Environment Agency. They consider that successful application of fish passes in tidal estuarine environments is as yet unproven. The Environment Agency has therefore taken a "precautionary principle" until there is further proof that such fish passes work. In this respect they have worked to the principle that they will not endorse the scheme until the promoter has "proven" that the fish pass will work. This is carried out through their regulatory consents and approvals process. Unfortunately for the promoter it is difficult to prove this until after construction. Therefore promoters have taken on the liability of proving the design in its final situation. This necessitates both pre- and post-construction comprehensive monitoring. Promoters should be fully aware of this liability during the planning stage. However, as precedents are set this liability will become easier to assess.

5.5.3
Planning Policy Guidance and Other Legislative Instruments

It is important that the proposed barrage scheme complies with government policies, and strategies such as the planning procedure will examine the scheme in this context. The Government, Environment Agency, local authorities and many other organisations are developing procedures to achieve effective coastal zone management. The Severn Estuary Strategy is an example of a collaborative venture which will influence the land and water environment development.

The UK Government's objectives for nature conservation, including the means by which UK obligations under international conventions and European and national law are to be met, are detailed in *Planning Policy Guidance Note, No 9, Nature Conservation* (DoE 1994). With regard to protection of species and habitats a variety of legislative instruments are of significance:

UK
- Schedules 5 and 8 of the UK Wildlife and Countryside Act 1981 list animal and plant species which are strictly protected under the Act.
- Terrestrial sites of significance for nature conservation within England and Wales are notified by English Nature and the Countryside Council for Wales as Sites of Special Scientific Interest (SSSI) under the Wildlife and Countryside Act 1991.
- Marine sites of conservation significance within the UK are designated as Marine Nature Reserves (MNR) under the UK Wildlife and Countryside Act 1981.

European
- Annexes IV and V of the EC Habitats Directive list animal and plant species in the European Union in need of protection. Part III of the Conservation (National Habitats, and Conservation) Regulations (HMSO 1994) provide for the protection of these species occurring in the UK.
- Wetland sites of international significance such as waterfowl habitat are designated as RAMSAR sites under the "RAMSAR" Convention on Wetlands of International Importance.
- Additional designations apply to areas protected as European sites under EU Directives implemented through the Conservation (Natural Habitats and Conservation) Regulations 1994.
- Special Protection Areas (SPA) are designated under the Birds Directive. Many estuaries and coastal areas come into this category.
- Special Areas of Conservation (SAC) are designated under the Habitats Directive as are European Marine Sites.

After an assessment of a scheme's likely impact on the conservation objectives of any of the variously designated sites listed above, development can only be permitted where "in the absence of alternative solutions it must be carried out for imperative reasons of overriding public interest – including those of a social and economic nature". If a development is allowed to proceed on such a basis all compensating measures must be taken to ensure that the overall coherence of the European network of SPAs and SACs – known as "Natura 2000" – is maintained.

5.5.4
The Habitats Directive

The Habitats Directive has significant implications for future barrage schemes and it is therefore appropriate to discuss this particular directive in more detail. These implications were examined by Huggett (1996). The key element of this Directive, in the context of barrage construction, is that pertaining to the habitat protection requirements and in particular that of "favourable conservation status". This is being achieved with the creation of the European-wide network of designated areas, called the Natura 2000 Network, which comprises Special Protection Areas (SPA) and the Special Areas of Conservation (SAC). Fundamental to barrage development is the requirement for member states to ensure that the habitats within the SACs do not deteriorate and that species are not exposed to significant disturbance.

This Directive sets the agenda for the decision makers. The procedure includes undertaking an assessment of whether there will be "adverse effects on the integrity of the site" due to the (barrage) construction. If there will be, then consent must be refused unless alternative less damaging solutions can be found or there is imperative overriding public interest for giving consent. In this case compensatory measures should be undertaken.

The integrity of the site is defined in the *UK Planning Policy Guidance Note, No. 9, Nature Conservation*, as the coherence of its ecological structure and function, across the whole area, that enables it to sustain the habitat, complex of habitats and/or levels of populations for which it was classified. It is not possible in this chapter to examine the full implications of this Directive. At first glance the Directive precludes changed environments which would be generated by many barrages. However, many of the implications are still to be tried and tested. Huggett concluded:

> "Implementation of the Habitats Directive, along with the Birds Directive, is complex. Both raise a number of questions to which there are currently no definitive answers. Indeed, with respect to some questions it is likely that agreement on the answers and correct interpretation of the law will not be immediately possible. In such cases, it may be necessary to seek clarification from the European Commission through the European Court. Similarly, the implications of a barrage development are not necessarily certain or clear cut. Reducing uncertainty and the degree to which the precautionary principle must be advocated will be needed before some tests of the Directive can be adequately passed.
>
> However, whilst these problems are significant, they should not be insurmountable. The Habitats Directive should not be seen as a barrier to barrages. Indeed, it may be that some barriers may be needed in the future to actually protect the nature conservation interests of the SPAs and SACs. What the Habitats Directive does provide is a framework within which to balance the economic and social needs of a Member State with the European importance of the site and the Natura 2000 Network as a whole."

5.5.5
Recreation and the Amenity Plan

Environmental policy must also take account of recreational needs as well as the ecological needs. Provision for recreation facilities is usually one of the main criteria in the planning of an amenity barrage scheme. Facilities may be provided for the following recreation types:

- Sailing
- Windsurfing

- Canoeing
- Fishing
- Leisure cruising
- Waterskiing
- Rowing
- Visitors (casual viewers and spectators)

A primary aid in the provision of recreation facilities can be the development of a recreation and amenity plan. This can serve as a useful source of information for the developer as it will identify the need for leisure facilities and also conservation areas that need to be protected. Consultation with interested parties/stakeholders at an early stage is paramount in order to optimise the leisure resource and so those interested parties may also "own" the plan.

5.6
Conclusion

Estuaries have a high conservation, environmental and amenity value and are also a very limited natural resource. Barrages have the potential to have a considerable impact on them. It is essential therefore that, when decisions are made, all environmental costs are fully taken into account. It is also essential that alternative options are fully considered as a proper evaluation may reveal that they achieve the objectives with lesser environmental impact. Any barrage proposal will need to thoroughly demonstrate that it conforms to the principles of sustainable development, namely that the development will not compromise the needs of future generations. This includes not passing unmanageable or costly liabilities on to those who operate the barrage in future years.

It is vital that all concerned appreciate that the current understanding of the impact of estuarine barrages and the associated mitigation measures is very poor. Therefore in some cases there may be a need to apply the precautionary principle with the result that parts or indeed the whole development are not progressed until the impacts are fully understood. As briefly discussed in this chapter the guidelines will provide a detailed review of both the *issues* relating to barrage planning design construction and maintenance and the *environmental policy* framework in which the resolution of these issues is undertaken. These guidelines may form the agenda for clarifying the environmental policy for barrages and thereby establishing an integrated management philosophy. It is hoped that the guidelines will act as a reference for many involved in such a scheme. This goes some way towards the interdisciplinary training that is required for an integrated management approach.

Barrage projects like any other large-scale projects in the coastal environment must take full account of all environmental, social and cultural effects and economic cost and these should be assessed using the best available technology.

Acknowledgements. It should be noted that this project is still underway and that the views expressed by the authors are not necessarily those of HR Wallingford, the Department of the Environment or the Environment Agency. The authors wish to acknowledge the contributions of the following members of HR staff as follows: Project manager: Mr. T.N. Burt; concepts, planning and legislation: Mr. I.C. Cruickshank; structural concept and design: Mr. J.D. Simm, Mr. C.J. Pyne, Mr. I.C. Cruickshank; hydrody-namics: Mr. N.V.M. Odd; morphology: Dr. R. Bettess; flooding: Dr. P.G. Samuels; groundwater: Dr. C.E. Reeve; water quality: Mrs. J.M. Maskell; navigation: Dr. I. Mc Callum (HR Mardyn); waves: Mr. P.J. Beresford; fisheries and conservation: Dr. A.S. Nottage; water-related risk: Dr. P.G. Samuels, Mr. I.C. Meadowcroft; project co-ordination and editorial assistance: Mr. M.A. Littlewood.

The authors also acknowledge the contributions of Mr. P. Hunter, Dr. P. Mason, Mr. S. de Turberville and Mr. N. Pope of the staff of Sir Alexander Gibb and Partners to the chapter on Structural design and for review comments on other chapters in the guidelines.

The research project has been carried out under the guidance of a Steering Com-mittee set up by the Environment Agency. The advice and information provided by the members in the preparation of the guidelines is gratefully acknowledged: Dr. A.W.G. Rees (Chairman), Mr. Peter Gough, Mr. Kevin Thomas, Mr. Wayne Davies, Mr. Richard Howell, Mr. John Lambert, Mr. David Mee, Dr. Dafydd Evans, Mr. Paul Tullett, Dr. Steve Axford, Mr. David Wilkes, Mr. Tim Barrett and Mr. Paul Varallo.

Bibliography and References

HR Wallingford (1997) Guidelines for the design and operation of estuarine barrages. Final internal draft expected winter 1997

Burt TN, Watts J (1996) (Eds) Barrages – engineering design and environmental impacts. Int. Conf. Cardiff Sept. 1996, John Wiley & Sons

Burt TN, Cruickshank IC (1996) In: Burt TN, Watts J (1996) (eds) Tidal barrages – learning from expe-rience. Barrages, engineering design and environmental impacts. Proc. Int. Conf. Cardiff Sept 1996, Wiley, Chichester, pp. 3–17

CIRIA (1994) Environmental Assessment. Construction Industry Research and Information Associa-tion, Special Publication No. 96. CIRIA, London

DoE (1988) Environmental Assessment. Department of the Environment Circular 15/88, HMSO

DoE (1989) Environmental Assessment – a guide to the Procedures. Department of the Environment, HMSO

DoE (1992) Government Circular, DOE Circular 30/9, WO Circular, 68/92, On development and flood risk. HMSO

DoE (1992) PPG 20, Planning Policy Guidance: coastal planning. HMSO

DoE (1994) PPG 9, Planning Policy Guidance: nature conservation. HMSO

DoE (1994) PPG 23, Planning Policy Guidance: planning pollution and control HMSO

DoE (1994) Evaluation of environmental information for planning projects – a good practice guide. HMSO

DoE (1994) Environmental Assessment, amendment of regulations. Department of the Environment Circular 7/94, HMSO

DoE (1995) A guide to risk assessment and risk management for environmental protection. HMSO, Lon-don

DoE (1995) Permitted development and environmental assessment. Department of the Environment Circular 3/95, HMSO

DoE (1995) The Habitats Directive – how it will apply in Great Britain. Department of the Environment, HMSO

DoE (1996) Review of the potential effects of climate change in the United Kingdom. Department of the Environment, HMSO, London

HMSO (1992) Transport and Works Act 1992, a guide to Procedures for obtaining Orders relating to transport systems, inland waterways and works interfering with rights of navigation. HMSO, Lon-don

HMSO (1994) Sustainable development, the UK strategy. HMSO, London 1994

HMSO (1994) Minerals Planning Guidance: Note 6. HMSO, London

HMSO (1994) Part III of the Conservation (National Habits and Conservation) Regulations, HMSO, London 1994

Huggett D (1996) The Habitats Directive: a barrier to barrages? Paper presented at the International Conference on Barrages, Cardiff, UK, September 1996

CHAPTER 6

Building Bridges to Science:
Making Coastal Science Better Understood

Carolyn Heeps · Vincent May

6.1
Introduction

Chapter 17 of Agenda 21 sets out major guiding principles for coastal zone management in the context of sustainable development whilst Chap. 36 emphasises that education "provides access for concerned individuals, groups and organisations to relevant information and opportunities for consultation and participation in planning and decision making at appropriate levels" (Quarrie 1992). Facilitation and implementation of any effective coastal management process therefore demands improved links and knowledge of natural environmental and socioeconomic factors through effective interpretation, presentation and dissemination of data and information. Amongst governments and businesses alike there has been a growing recognition that understanding of scientific principles and methods is poorly understood. There has been only very limited investigation of the public understanding of coastal and marine science, although there are growing numbers of visitor centres providing environmental interpretation to the public. Such interpretation should, however, do more than merely inform, as it can be used to change attitudes and behaviour and to mould opinion. Despite increasing this public exposure to the coastal and marine world, it is evident from the debate about both large- and small-scale construction in the coastal zone that there is very limited understanding of fundamental scientific, engineering and statistical principles which support the arguments for or against development. Similarly, quasi-scientific information is sometimes used to make political statements rather than scientific argument. This chapter discusses the problems of public understanding of science (especially in connection with project development in the coastal zone) and describes and evaluates a selection of projects in Dorset, England, which are building bridges to coastal and marine science.

6.2
Development, Science and Public Understanding

Most management problems occur within jurisdictional waters, and many coastal communities recognise the need for the sustainable use of both living and non-living resources. Informal surveys, however, reveal that the general public list the more immediate issues of pollution (especially water quality, clean beaches and loss of amenity value), overexploitation of marine resources (especially fisheries), coastal erosion and flooding, sea level rise and habitat loss as major issues of concern which urgently need to be addressed by coastal management. In contrast, the scientific concepts behind these issues are commonly poorly understood and the media often provides the interface between industrialists, environmental scientists and the public. The marine environment is still an alien world, not just to the general public but also to many sectoral interest groups. Sustainable development requires an integrated and more holistic approach to coastal management, based upon a deeper understanding of physical forcing functions and ecosystem processes. Unfortunately, unless individuals or interest groups are directly affected by an issue they are reluctant to learn more about it, voice their concerns or become actively involved in the management process.

Another aspect of the debate is the role of technology in both helping to solve and sometimes adding to problems. In a global context many of the products and services of the marine sector relate to large-scale construction (e.g. oil and gas pipelines, port facilities, bridges, coastal and flood defence schemes), much of which takes place in shallow coastal waters where environmental impacts are most pronounced. The increase in subsea operations, emerging issues of decommissioning and waste disposal, changes in marine transport (particularly to serve the leisure market) and developments in aquaculture all require improved resource and coastal management (Office of Science and Technology 1997). With population migration to the coast continuing to escalate, there will be added pressure and impact upon coastal and marine resources in the future.

If effective management requires advice and decision making based upon best available information, then the emerging practice of coastal zone management should be viewed as an opportunity to provide a longer-term vision and actively engage all sectors of the community in data collection, assimilation and dissemination. The UK Marine Foresight panel (OST 1997) emphasised that such innovation and technological change must be matched by a development dimension in the provision of marine education and training at all levels. The new skills required for industry, professional institutions, trade associations and academia to compete within the global market place are rapidly changing: such bodies need to attract and encourage new human resources into the marine science and technology sector. This is not intended to imply that there is a greater need for traditional marine scientists, but rather that the sector has poor links to the social sciences. Moreover, it has not been effective in encouraging public understanding of the nature and processes of the marine sector. This can be achieved through the stimulation of the public understanding of the marine environment and its value to future heritage. The task – to improve everyone's knowledge and appreciation of the oceans – is not a simple one.

This arises at least in part because there are considerable barriers to people's understanding and use of scientific information. The ways in which scientists work and the extent to which their results approximate to the truth are poorly understood by developers, decision-makers, the media and the public. The progress of science as a powerful tool for development has depended upon its practice of doubting theories, even those on which action itself is based (Haldane 1927). Increased public participation in the debate about large-scale coastal construction challenges coastal scientists and developers to move from a position in which scientific knowledge is the preserve of "experts" to a shared fund of knowledge accessible to all who use and live in the coastal zone.

6.3
Constraints on Public Understanding of Coastal Science

Five key issues affect public understanding of coastal and marine science: existing levels of knowledge and the uncertainties surrounding the data itself, the decision-making processes for development projects, the nature of environmental science communication, the role of scientists as translators, and public attitudes to science and scientists. First, the actual levels of understanding of environmental principles and processes amongst the public, decision-makers and practitioners in cognate fields of study are complicated by the extent to which existing knowledge is uncertain or contains a heavy reliance on statements of probability. Despite public involvement in gambling, the concept of probability applied to scientific data is poorly understood and leads to serious misunderstandings. Two particular areas of difficulty are the use of return periods and understanding of relative magnitudes, for example in the case of sea level rise as compared to extreme wave heights. In order for decision-making to be carried out effectively, there is often a need to provide simple correct explanations of complex and changing phenomena, together with indications of the level of certainty which can be given to the information. However, the scientific methods themselves are uncertain because of the partial nature of data and the acceptance of probability. It is very rare for scientists to provide unequivocal answers to questions – that is not the nature of science. There is an urgent need for science to be presented clearly (Pollock and Steven 1997), whilst recognising that development decisions will be made, however partial the scientific evidence. Use of the precautionary principle is often supported, but this itself is often poorly explained or open to different interpretations.

Second, there are problems with the decision-making process for scientists and others dealing with development projects. Decisions are characteristically made (in Western democracies at least) by committees of elected representatives (who are often not experts) advised by professional officers. The latter may have the task of interpreting detailed technical reports produced by consultants. The same information chain often exists in project development in developing countries where there are different traditions of decision-making. Knowledge may be exchanged by oral means with an holistic rather than sectoral view. Environmental knowledge may be held by local communities (for example artisinal fishermen) but not in a form familiar or acceptable to many scientists.

Third, there is a need for greatly improved scientific communication, especially as many large-scale projects involve many disciplines. As Berridge (quoted in Wolpert and Richards 1997) has pointed out, the task of good science is to make connections between different ideas, and different disciplines. This requires the construction of *"social bridges"*, because public attitudes are complicated by the need for scientists to "translate" their findings even to other scientists (Uglow 1996). As a result, scientific information is often communicated in familiar language which lacks precision, may depend upon the use of analogies and is full of unscientific resonances (Greenlaw 1996). At its worst, the public has an image of "irresponsible creativity by science that refuses to think of consequences" (Belton 1996, p 263).

Fourth, the recent increase in scientists as media personalities and the growth of popular scientific writing may have made science more interesting, but it has not removed this inherent mistrust of scientists. As scientific certainties are shown to be fallible or at worst even dishonest, people show a growing tendency to be attracted to the non-scientific or untested explanations of phenomena, as witnessed by the growth of interest in para-normal phenomena. The search is not so much for understanding, but for explanations. The problem seems to arise from a confusion between science, whose role is to tell us what will happen and why, and policy which determines what should happen. Scientists and non-scientists alike are involved in the shaping of policy, even though many scientists cling to a view that they should remain uninvolved. However, Durant (1997) has indicated that when dealing with complex environmental issues, the judgement of non-scientists is often very similar to that of scientists. This is both worrying and helpful, for it suggests that people's innate understanding of their environment and how it works is strong, and that scientific analysis of the problems often supports this understanding. It challenges scientists to be less arrogant about their expertise!

Fifth, difficulties arise from the way in which scientific information is promulgated. The public have become confused because science is commonly used by politicians, media and NGOs to support particular points-of-view and so can be viewed as propaganda. Furthermore, in common law jurisdictions such as in the USA and Britain the information can be challenged in courts of law and may be shown to be incomplete or open to debate. This is especially so when predictions have been made, the magnitude of problems has been overemphasised and in effect the community has "cried wolf". The environmental movement has been particularly prone to produce forecasts of forthcoming doom or ecological collapse. For example, oil spillages attract massive media attention, often accompanied by predictions of ecological disaster, and yet some coastal ecosystems have appeared to be remarkably resilient. The example of the *Braer* oil spill off the coast of Shetland in 1992 demonstrates that apparent recovery within relatively short time scales can occur. Furthermore, some NGOs have become more business-like, their survival depending on maintenance of their public image. Cynics might argue that some make the worst of problems in order to ensure their survival rather than presenting solutions to the issue. There is often a perceived need to act even when scientific information is very limited. Action is taken on a cause-and-effect basis without much attention to either the complexity of systems, local conditions or the effects of public reaction. As a result, short-term reactions to a problem can produce a *"Verschlimmbesserung"* effect, i.e. the problem is worsened by action which it was believed would make the situation better.

Academia is often considered to be in the most prominent position to improve public understanding of science but historically does not have a good track record of community outreach. Although there is an element of *trust in the science* this is not always matched by the public *trust of the scientists*. What is important is that the public is brought into the discussion and debate that scientists are able to generate. Much information aimed at the public about environmental issues is delivered by the media who may gather and manipulate their information from a variety of disparate and often conflicting sources. At a more local level visitor centres play a vital role in increasing the public understanding of science because it is here that the public often have their first structured introduction to the marine environment. Thus, the general public tend to rely upon not only the media but also public environmental organisations (such as National Parks or local authority environmental services), local wildlife organisations, or NGOs/pressure groups running campaigns, for information. Industry turns to its own experts and advisors, professional institutions or to links with academia. Consequently there is still a need to improve vertical and horizontal communication and exchange of scientific information between industry and engineers, coastal scientists, decision-makers, the media and coastal user groups.

6.4
Making Coastal Science Better Understood

In the context of coastal management the provision of education and interpretation provides an opportunity to: 1) explain and promote the relevance of science and the scientific process in environmental decision-making; and 2) improve the dialogue between the range of organisations involved in decision-making. The social role of science in coastal zone management cannot be underestimated, particularly in relation to health and safety issues such as water quality and levels of contaminants in food resources. However, many concerned individuals and interest groups do not want the technical information or raw data sets, preferring to have the science translated and presented into a more user-friendly format. With information technology playing such a significant role within the industrial marine sector and scientific community, environmental educators need to seize the opportunities currently being presented by electronic transfer of data and multimedia applications to get information across to increasing numbers of visitors and home computer users. However, the scientific process must not be compromised. To achieve this, it is necessary to *"access the inaccessible"*, *"provide a public window to the marine environment"* and *"share the secrets of the sea"*.

6.5
Accessing the Inaccessible

The marine environment is for most people inaccessible, both physically and intellectually. It is dangerous and requires specialised equipment to go below the surface. It is also inaccessible to most people scientifically, because its characteristics are poorly understood. The growth in wildlife tourism and increasingly large numbers of visitors to the coast have presented providers of educational and interpretative resources with many new opportunities and challenges. In England the establishment of Voluntary Marine Reserves has played an important role in both raising awareness of the marine environment and encouraging capacity-building amongst the variety of stakeholders at a site-based level. The phenomenal rise in communications technology and multimedia provision (personal computers, CD-ROM and Internet accessibility) has given visitor centres an opportunity to provide innovative interpretation and learning opportunities which complement more traditional interpretative techniques such as guided walks and practical hands-on activities to facilitate discovery, observation, investigation and research.

In the UK the technological approach to interpretation is being used to good effect as a means of capturing the imagination and interest of visitors to the Dorset coast. Here, marine interpreters are providing new education and interpretation initiatives to inform the visiting public and school groups as well as encouraging community involvement in management tasks such as data collection. Innovative techniques such as seabed hydrophones, clifftop and underwater cameras are being used to observe and monitor coastal habitats and wildlife and transmit "live" sounds and pictures into visitor centres where the information is then interpreted. Such remote techniques enable large numbers of visitors, including the disabled, to have access to coastal habitats and wildlife with minimum damage and disturbance. Advances in multimedia provision within interpretation centres are actively engaging the public in "hands-on" science discovery which has the added benefit of encouraging them to become more actively involved at specific sites through volunteer or "Friends" groups. At a more strategic level educational strategies and networking initiatives are also developing; marine educators within Dorset have established their own voluntary network, called Dorset Coastlink, in order to take marine interpretation and education into the next millennium. However, their continued success will depend heavily upon adequate financial resourcing which has yet to be secured.

6.6
Providing the "Public Window to the Marine Environment"

Situated midway along the English Channel coast, the inshore waters and coastal geology of Dorset are internationally renowned. The coast is a popular destination for tourists, recreational users and educational groups. This south-western area of the English Channel contains many marine species at the northern limits of their geographical range; it is also one of the busiest shipping areas in the world and an area that is extensively fished. More recently, oil and gas exploration has been carried out in close proximity to the shore, raising environmental concerns along the Dorset coast which is a proposed World Heritage Site. Yet most of its splendour is hidden from the view of the many people that work these waters or rely on them for leisure and recreation. To remedy this, a number of visitor centres along the Dorset coast are making the marine environment accessible to all. By investing in high technology equipment these centres are capturing the imagination of visitors by providing the "public window to the marine environment". The key sites are (Fig. 6.1): Durlston Marine Research Area, Kimmeridge Bay Voluntary Marine Nature Reserve (Purbeck Marine Wildlife Reserve), Lulworth Cove Heritage Centre, Chesil Bank and Fleet Nature Reserve and the Charmouth Heritage Centre. Table 6.1 summarises the techniques used to present marine information at each site. The methods used to facilitate improvements in the public understanding of science at three of those sites (Durlston, Kimmeridge and Charmouth) are described below.

A Marine Research Area and Coastwatch Project is located at Durlston Country Park. Sheer limestone cliffs prevent access to the shoreline, but the cliffs and the sea around the headland have become accessible using modern technology. The colony of guillemots (*Uria aalge*) at Durlston Country Park is normally hidden from the direct view of visitors walking the clifftop path. Until recently only boat passengers had an opportunity to view them at relatively close quarters, but not without some level of disturbance. In 1993 a real-time video camera was installed on the cliff enabling live pictures of the guillemot colony to be transmitted back to a TV screen in the visitor centre. Visitors are able to witness the "private lives" of parent birds and their young throughout the spring/ summer breeding season. The clarity of the pictures and the ability to view the colony without disturbance provide a superb opportunity not only to enhance visitor enjoyment but also to help manage visitors more safely and effectively. Introductory sessions with visiting parties before they explore the park and its clifftop walks allow centre staff to explore a wide range of coastal topics using the camera. Live pictures from the cliff are used to explain ecological and conservation issues, whilst pictures of the sea and marine traffic including ferries, military vessels, cargo carriers, fishing boats, lifeboats and recreational craft illustrate the economic importance of the sea as well as the hostile nature of coastal waters in storm conditions.

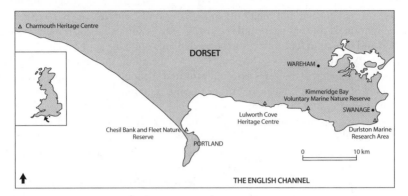

Fig. 6.1. Centres participating in Dorset Coastlink. [Based on 1973 Ordnance Survey 1:250000 map with permission of the controller of Her Majesty's Stationary Office © Crown Copyright]

Table 6.1. Interpretative Themes and Techniques at Dorset Coastlink Centres

Site	Key Characteristics	Key Interpretative Techniques/Projects	Public Participation
Durlston Marine Research Area	Sea cliffs – guillemot colony – geology Regular dolphin sightings Subtidal habitats Water-based recreation Commercial marine traffic	Clifftop camera linked to the visitor centre Fixed seabed hydrophone linked to visitor centre Dorset coast bulletin board Coastwatch Project SEAHEAR project for partially sighted Seasearch – underwater surveys	Friends of Durlston Dolphin sightings scheme Volunteer diver surveys
Kimmeridge Bay VMNR (Voluntary Marine Nature Reserve)	Extensive intertidal rock ledges Rocky shore ecology Geology and fossils Recreation and education Slipway access to the bay	Aquarium tanks in interpretation hut Rockpool temperature loggers linked to PC Remote underwater video camera Guided walks and snorkelling Limpet Protection Zone	Friends of Kimmeridge Volunteer wardens Underwater surveys
Lulworth Core Heritage Centre	Geology and geomorphology Slipway access	Heritage Centre displays Audiovisual presentation Education room Guided walks	
Chesil Bank and Fleet Nature Reserve	Chesil Beach – shingle bank Fleet saline lagoon Little tern colony Portland Harbour	Interaktive PC display Aquarium tank Guided walks Glass-bottomed boat on Fleet	Friends of Chesil Beach centre Pollution response exercises
Charmouth Heritage Centre	Jurassic coast Fossils Landslides Subtidal habitats	Audiovisual room – Jurassic theatre Fossil touch table Computer fossil identifier Guided fossil hunts Marine Awareness Project Aquarium tank CD-ROM marine life photographs	Friends of Charmouth

The main means by which the submarine world communicates is by sound. Light transmission is very poor – even in the clearest waters the human visual range is less than 30 m and visibility is often less than 5 m. Organisms in the sea use sight as a means of close-up recognition, but many organisms respond to variations in pressure and marine mammals communicate and locate themselves sonically. Underwater communication and detection for military purposes depends upon the application of sound as does seabed mapping and investigation. However, the public understanding of the nature and use of sound in the sea is extremely limited. Although many people have heard the sounds of dolphins, whales and other marine mammals and know that they use sound to navigate, hunt for food and communicate, very few have seen the seabed portrayed on sonographs. An informal survey of scientists, engineers as well as the wider public showed that whilst many are aware that the speed of sound in air is about 340 m s^{-1} (95% of respondents), the same percentage did not know if sound travels faster in air or water. Even fewer knew that the average speed of sound in water is about 1540 m s^{-1}. In a scientific group, including individuals with degrees in physics, knowledge of the speed of sound in water was either absent or faulty. In order to overcome this lack of knowledge, a number of related projects have been developed which rely primarily on most visitors' fascination with dolphins.

A hydrophone was sited in 12 m water 400 m offshore in Durlston Bay in 1993. Underwater sounds in the Bay are cabled to the centre where they can be recorded for future analysis and transmitted as part of an interpretative display on underwater acoustics. Filters are used to enhance dolphin sounds for human hearing. Marine sounds within frequencies 10 to 20 kHz are recorded, but filtering can detect sources within an envelope of 200 kHz. The hydrophone picks up not only the echo location of the dolphins as they map out their environment, search for food and communicate with each other, but also a wide range of ambient sounds, including those made by other marine animals, passing ships, seismic exploration, power boats, personal watercraft, pile driving, even land-based quarrying, thereby giving a true indication of noise levels in coastal waters. There appear to be two distinct sounds associated with cetacean activity. One is a comparatively slow click rate, which usually coincides with sightings of one or more bottlenose dolphins (*Tursiops truncatus*) just offshore. The second sound, a much faster click rate with high whistles, is present most of the time. These may be produced by common dolphins (*Delphinus delphis*) or harbour porpoise (*Phocoena phocoena*). There are many other sounds, including a continuous crackling which may be snapping shrimps or mussels opening and closing their shells. Unidentified explosions may come from local quarries and naval sonars have been picked up from frigates operating over 20 km away. High frequency fast repetition clicks have been recorded for about 80% of the time since July 1994. High frequencies propagate less well in water so the frequent recording of these clicks implies that the sound source was generally within 500 m of the hydro-phone. A dolphin sightings scheme, using volunteer observers, has highlighted the importance of the waters off Durlston for feeding and migrating bottlenose dolphins. Even today, many first-time visitors to the centre do not believe that dolphins can be seen off the Dorset coast until they encounter the displays within the centre. Using this new knowledge, visitors are then encouraged to consider noise pollution in the sea as an environmental issue, the potential disturbance on wildlife from some human coastal activities and how the information collected can actually aid managers in their management tasks. In addition to the dolphin sightings scheme, a comprehensive photo-identification survey is providing valuable information with respect to dolphin populations in this section of the English Channel. Collectively these projects are known as Coastwatch which "aims to raise awareness of the coastal waters off Durlston Head".

Research and interpretation projects using the sounds being recorded continue to be developed; SEAHEAR is one special project developed by the School of Conservation Sciences at Bournemouth University. A multimedia, computer-based system introduces visitors to the complex science of acoustics, with particular emphasis on underwater acoustics. The project uses a wide variety of information and images related to the use of acoustics and is also addressing the provision of interpretation for the blind and visually impaired – just as dolphins use sound to communicate and "see" underwater, sound can also be used to explain the marine environment to those with sight difficulties.

The scientific projects that make up the Coastwatch project rely heavily on the input of volunteers – the "Friends of Durlston" group. These Friends get involved in a wide variety of tasks including the dolphin sightings scheme, manning the interpretation centre, and most recently, carrying out underwater biological surveys in the Bay. The community partnership approach has developed beyond the centre; local commercial boat operators are being encouraged to incorporate more interpretation into their short cruises along the coast and the statutory Environment Agency has helped to map the Marine Research Area using side scan sonar.

Established in 1978, the Purbeck Marine Wildlife Reserve in Kimmeridge Bay has no legal status but is one of the increasing number of Voluntary Marine Nature Reserves (or Conservation Areas) along the coast. It is a prime example of a successful site-based conservation initiative which depends on the voluntary co-operation of local landowners, conservation, recreation and interest groups and regular users. The site receives thousands of visitors; school parties, divers, anglers, windsurfers, surfers, geologists and family groups all enjoy what the bay has to offer. A small interpretation centre, open during the summer season, houses aquarium tanks representing the different habitats found in the bay and the full-time warden also leads school visits, guided walks, rock pool rambles and organises special events such as underwater photography competitions. The most recent additions to the interpretation provision include instrumentation to monitor environmental conditions in rock pools and a remote underwater video camera which is used to bring pictures "live from Kimmeridge Bay" to shore visitors.

Innovative approaches to education do not have to rely on multimedia advances and computers: at Kimmeridge the establishment of a "Limpet Protection Zone" (LPZ) has been very successful, and amusing! Increasing numbers of visitors to the shore over the years has resulted in unnecessary removal of limpets (e.g. *Patella* spp.) so volunteers patrol the beach informing visitors about the significance of the limpet to rocky shore ecology. The initiative has received much media attention during the summer holiday season at both a local and national level.

World renowned for its fossils, Charmouth Heritage Centre has naturally focused on the geology of Dorset's internationally famous Jurassic coast. However, the coastal and marine component of interpretation is not being overlooked; a Marine Awareness Project has been initiated in conjunction with the establishment of a voluntary marine conservation area. Here, visitors have access to an increasing range of multimedia techniques in addition to the traditional fossil hunts with the warden. Fossils such as ammonites are being used to explain to visitors what the seas were like during the Jurassic period, so that visitors can begin to compare it with life in the sea today. A computer-based fossil identifier has been developed to encourage visitors to identify the fossils they have found themselves rather than always relying on the availability of a member of staff or centre volunteer. A small aquarium provides a living display about the rich marine life to be found in the waters of Lyme Bay today and is accompanied by a CD photo display which visitors can work through at their own pace. The computer-based projects have been developed with undergraduate computing students from Bournemouth University, thus demonstrating that productive relationships can be nurtured between universities and the public as well as providing important opportunities for students to work on "live" projects.

6.7
"Sharing the Secrets of the Sea" – Dorset Coastlink – A Case Study in Networking and Promoting Public Understanding of Coastal Science

Individual centres, however, can benefit from sharing the information. It also helps manage visitors to a very popular coastline. In 1994, therefore, educators from each of the Dorset sites met to discuss the range of educational and interpretation resources available. This informal meeting resulted in the formation of a unique voluntary networking group called Dorset Coastlink, bringing together the individuals and organisations involved in providing education and interpretation. With the shared aim "to raise awareness of Dorset's marine and coastal environments", the network provides a much more coordinated approach to the protection and promotion of Dorset's coastal waters and in so doing has taken a national lead. A countywide interpretation and education strategy is currently being developed which will make use of advances in electronic transfer of data – not only between sites themselves but also to other organisations such as schools, universities and tourist information centres, many of which may be outside the county. It is hoped that this technology will allow potential visitors to become better informed before their visit and plan accordingly. Alternatively, education groups will be able to study the Dorset coast throughout the year rather than during a 1-week field visit, or without having to visit at all! Each centre within the Dorset Coastlink network is able to develop its own interpretation, education and research projects, but is doing so within a local framework rather than as an isolated site-based approach. The network facilitates the exchange of information and sharing of expertise and skills within the group and each centre's community; all have a growing army of volunteers who receive training for interpretative and research projects.

Another good example of the benefits of a more strategic approach to the provision of interpretative opportunities for visitors is Marine Week, an annual event organised by the Charmouth Heritage Centre. Marine Week encourages the general public to take a fresh look at the marine environment, with activities ranging from boat trips and using remote underwater cameras, to fossil walks and slide shows at various locations within the locality. Many different organisations are involved, ranging from the local NGOs, businesses and industry through to trade associations, thereby facilitating a closer working relationship between the different interest groups. Being part of the Coastlink network means that the other centres along the Dorset coast can schedule their own activities and events around this special week to avoid clashes or to help promote the Charmouth events.

In Dorset, stimulating and innovative education and interpretation coupled with a strong network approach is:

- Providing visitors with a quality experience
- Improving the public understanding of science
- Enabling effective communication, information exchange and dissemination, and the sharing of resources between educators
- Strengthening institutional links
- Establishing baseline monitoring programmes and data collection
- Actively involving the community
- Putting coastal issues on the local political agenda
- Gaining support for coastal management
- Advancing research
- Stimulating interest for similar projects elsewhere
- Enabling marine education and interpretation to be accepted as a discipline in its own right

The Dorset Coastlink project as a whole and its individual elements demonstrate that there are a number of essential requirements for success in improving public understanding of the coastal and marine environments:

- Strong practical support from voluntary, statutory and business organisations
- Adoption and deployment of the highest levels of modern information technology to supplement more personalised methods of interpretation
- Well-motivated and trained staff (including volunteers)
- Capturing of the public interest and imagination through their natural curiosity about the sea

Coastlink provides a model by which this can be achieved. Its success so far lies with the personalities involved in the network. It would be wrong, however, to suggest that the projects have been without problems. There is a continuing need to have adequate funding on a continuous basis. Grant-aid tends to be small and interpreters are expected to secure funding as well as develop projects. Many of the publicly funded environmental agencies only provide funding on a 50 : 50 basis and often regard their role as providing pump-priming rather than long-term maintenance of existing projects. This approach only brings about a further fragmentation of education and interpretation provision rather than a strategic, coordinated development approach. Furthermore, volunteers have to be trained and managed, but this area is fortunate in having a large constantly renewed population of retired people for whom such projects become a major part of their daily life. It is also probably Europe's most visited educational area and so there are specific problems of managing and informing large numbers of students of all ages.

6.8
Conclusions

At a time when public consultation and participation are being actively encouraged and considered to be a critical element in decision-making for marine and coastal environmental protection and management, it is vital that user and community groups have access to scientifically accurate information and data on which to make informed decisions. Better integration is needed between education, planning and local action. Interest groups are demanding a greater transparency within industrial and government agencies. Although improvements have been made in communication channels and the flow of information between these bodies through voluntary initiatives such as coastal fora, it is important that the scientific information is used appropriately and correctly.

There are thus four major issues for the improved understanding of processes in the coastal zone and the way they may be affected by development:

- Issues of scientific knowledge – what is known, how reliable this information is, and how certain are its predictions
- Communication – how is scientific information selected, communicated and used
- The place and training of translators or interpreters
- The provision of adequate long-term funding for education and interpretative projects

Public understanding of science and technology depends on the information provided by industry and the environmental NGOs as well as government agencies. Each agency has its own agenda, remit and legislative responsibility and public concerns often arise because the published scientific evidence differs from the perceptions of the people affected. Cultural indifference is not acceptable. There is a serious lack of competence in the social and economic understanding of coastal systems, but there is also a very poor level of communication of verifiable, rigorous information between the various sectors which work in the coastal zone.

Acknowledgements. We gratefully thank the UK Committee for the Public Understanding of Science (COPUS) for a development grant which supported part of the programme described above, Dorset County Council's Countryside Service at Durlston Country Park (and in particular its Coastwatch project for assistance in collecting underwater sounds) and the School of Conservation Sciences at Bournemouth University for continuing support for the SEAHEAR project.

References

Belton N. (1996) Candied porkers: British scorn of the scientific. In: Spufford F, Uglow J (eds) Cultural babbage: technology, time and invention. Faber and Faber, London, pp 240–265
Durant J (1996) Jury out in the cold on science. Times Higher, 19 January 1996:16
Greenlaw L (1996) Unstable regions: poetry and science. In: Spufford F, Uglow J (eds) Cultural babbage: technology, time and invention. Faber and Faber, London, pp 215–226
Haldane JBS (1927) Possible worlds and other essays. Chatto and Windus, London
Office of Science and Technology (1997) Progress through partnership: 16 (marine). HMSO, London
Pollock J, Steven D (1997) Don't patronise the public. New Scientist, 155 (2101):49
Quarrie J (1992) Earth Summit '92. The United Nations Conference on Environment and Development, Rio de Janeiro
Uglow J (1996) Introduction: possibility. In: Spufford F, Uglow J (eds) Cultural babbage: technology, time and invention, Faber and Faber, London, pp 1–23
Wolpert L, Richards A (1997) Passionate minds. Oxford University Press, Oxford

The Control and Steering Group for the Øresund Fixed Link (KSÖ) and the Environmental Control and Monitoring Programme

Monika Puch · Jon Larsen

7.1
The Control and Steering Group for the Øresund Fixed Link (KSÖ)

In 1991 the Danish and Swedish governments agreed on constructing a fixed link across Øresund between Copenhagen and Malmö. The Fixed Link consists of a combined twin track railway and a four-lane motorway. Travelling from Denmark to Sweden, the fixed Link's key elements are an artificial peninsula, an immersed tunnel under the Drogden navigation channel, an artificial island south of Saltholm and a bridge over the Flintrännan navigation channel. The construction works began in 1995 and the Fixed Link is expected to be completed in 2000.

After signing of the agreement between Denmark and Sweden the Fixed Link project has been subject to approvals in accordance with the legislation of the two countries.

By direction of the Swedish government, the Swedish Environmental Protection Agency (EPA), the County Administrative Board of Malmö (of Skåne since 1 January 1997) and the Department of Environmental and Public Health Protection of Malmö formed a Control and Steering Group for the Øresund Fixed Link (KSÖ) in September 1994 in order to co-ordinate the environmental supervision.

The Swedish Environmental Protection Agency is the central supervisory authority according to the Swedish Environment Protection Act. Together with the County Administrative Board it is also appointed to superintend the construction works according to the Swedish Water Act. The Department of Environmental and Public Health Protection of Malmö superintends the construction works according to the Environment Protection Act.

The main tasks of the Control and Steering Group are to:

- Co-ordinate the supervision of the Swedish authorities in connection with the construction of the Swedish part of the Øresund Fixed Link
- Act as co-ordinator between the supervision of the Danish and the Swedish authorities
- Act as the designer and implementator of a control programme for the whole Øresund Fixed Link
- Supervise the environmental control and monitoring programme and act as co-ordinator of the information from the programme.

In July 1995 the Swedish Water Rights Court gave its final approval for the construction of the Øresund Fixed Link. In the verdict, some more precise conditions for the Swedish part of the Øresund Fixed Link were ratified and, in addition to the above mentioned authorities, the following authorities were appointed to supervise the construction of the Øresund Fixed Link:

- The Swedish Geotechnical Institute – to superintend the monitoring of sediment spill
- The National Board of Fisheries – with reference to the fisheries
- The Swedish Maritime Administration – with reference to the shipping

Today the Control and Steering Group for the Øresund Fixed Link (KSÖ) is composed of:

- The Swedish Environmental Protection Agency with three representatives
- The County Administrative Board of Malmö with three representatives
- The Department of Environmental and Public Health Protection of Malmö with three representatives
- A secretariat with two employees who administer the daily work

The Swedish Geotechnical Institute is co-opted to the Control and Steering Group and participates with two representatives. The remaining authorities conduct their supervision independently but are also consulted through the Control and Steering Group when necessary (Fig. 17.1).

By direction of the Swedish government, the Swedish Environmental Protection Agency and the County Administrative Board of Malmö shall continuously inform the government about the fulfilment of the task. Among other things, this is done through semi-annual progress reports regarding the environmental consequences of the construction work on the Øresund Fixed Link's coast-to-coast installation, which are produced in co-operation with the Danish authorities [Ref. 1–4].

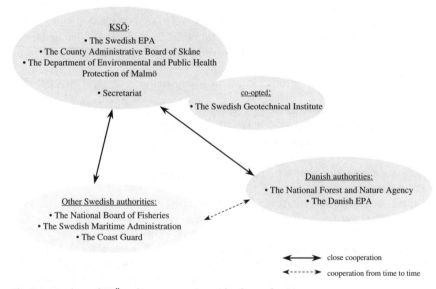

Fig. 7.1. Members of KSÖ and its co-operation with other authorities

7.2
The Environmental Control and Monitoring Programme

The Danish and Swedish authorities have defined a number of criteria for the accepted impact on the environment in Øresund and in the Baltic Sea [Ref. 5–10]. The construction of the Fixed Link must not change the water flow through Øresund, i.e. the salt and oxygen transport to the Baltic Sea must not be changed. Further the sediment spill from the construction works must not exceed 5% of the total dredged amount of approximately 7 million m³. Criteria have also been set up for water quality referring to nutrients, waste water, heavy metals, oxygen consumption, bathing water and for benthic fauna and vegetation referring to fish spawning and nursery grounds, breeding, resting and migrating birds, just to mention some of the criteria. An extensive control and monitoring programme is carried out in order to ensure that the environmental criteria are met.

The monitoring programme is primarily intended to record the state of the environment and forms the basis for counter measures if a construction activity has an effect which goes beyond what is expected. Following completion of construction, the monitoring programme is intended to record whether the temporary effects on the water environment decrease as assumed.

The responsibility for implementation of monitoring and control in connection with the construction work is divided between the contractors (spillage monitoring), Øresundskonsortiet (feedback monitoring for quick responses to environmental changes), and the Danish and Swedish authorities (monitoring of general effects). The authorities' monitoring programme is implemented by the Danish and Swedish environmental authorities with a view to investigating the long-term changes and to evaluating whether the observed effects during and after the construction work lie within the frameworks of the expected effects set out in the environmental impact analysis. Thus the monitoring period comprises the period until the completion of the construction work, but may be prolonged in order to document fulfilment of criteria on environmental re-establishment. The programme includes monitoring in the following sub-programmes:

- Water quality
- Benthic vegetation and benthic fauna
- Fish
- Birds
- Beaches and coasts

This chapter presents the structure of the authorities' environmental monitoring and control programmeThe results from the monitoring programme are described and evaluated in annual progress reports for each of the sub-programmes [Ref. 11–16], as well as in the semi-annual progress reports mentioned above (see Sect. 7.1).

7.2.1
Monitoring of Water Quality

The construction and operation for the Øresund Fixed Link must not lead to the release and redistribution of heavy metals, environmental hazardous substances or nutrients, or to oxygen consumption to an extent which creates negative ecological effects or leads to significantly increased concentrations of heavy metals or hazardous substances in animals and plants.

In order to monitor water quality, Secchi depth, nutrients, temperature, salinity, oxygen content and fluorescence are measured at each of the four stations in the programme (Fig. 7.2). The results of the weekly measurements are reported to the authorities within a week of the samples being taken. The results are supplemented by measurements carried out by regional and national Danish and Swedish authorities.

7.2.2
Monitoring of Benthic Vegetation and Benthic Fauna

It is generally accepted that there will be a temporary reduction in species composition, distribution and biomass of flora and fauna of up to 25% in the outer impact area, while permanent effects within the inner impact area of the alignment can be accepted.

7.2.2.1
Benthic Vegetation

The vegetation programme includes monitoring and control of the occurrence of pondweed (*Ruppia* sp.), eelgrass (*Zostera marina*) and sweet tangle (*Laminaria saccharina*). The programmes for pondweed and eelgrass are carried out once a year in August/September and include aerial photography and subsequent digital image processing of the distribution of the vegetation along the coast out to the 6-m-depth contour line (lower distribution limit for eelgrass in the Øresund) in the area from Aflandshage to Vedbæk, the coast of Saltholm, and the Swedish coast from Falsterbo to Landskrona. In addition to this analyses are made of the vegetation coverage and biomass of eelgrass and pondweed along transects within the outer impact area as well as the control area (Fig. 7.3 and 7.4). For eelgrass also the shoot density is measured. The monitoring of sweet tangle takes place at six localities in August/September using photography and surveying.

7.2.2.2
Benthic Fauna

The programme for benthic fauna includes monitoring of the shallow water fauna (<6 m depth) and deep-water fauna (>6 m depth), together with a separate programme for common mussel. Shallow-water fauna is collected at a small number of transects and stations during the spring, while a more extended sampling takes place in the autumn. The samples are analysed for the occurrence of species, number and biomass (Fig. 7.5).

Samples of deep-water fauna are collected annually in April/May at 17 stations with a view to determine biomass, number and size distribution. Common mussels are sampled annually in October/November at 48 stations. Coverage is assessed and biomass, number and size distribution determined (Fig. 7.6). The content of heavy metals in mussels at ten stations is also determined.

Fig. 7.2. Sampling stations for the water quality programme. Weekly samples are taken at stations 921, 431, 1728 and 441, and supplemented by samples from a number of Swedish regional stations and Danish county and municipal stations

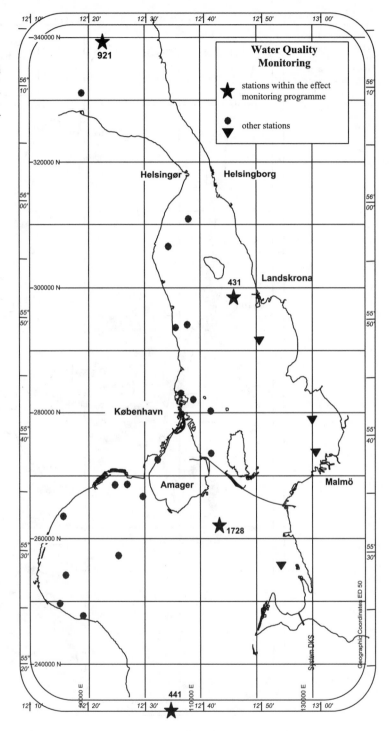

Fig. 7.3. Transects and stations for monitoring of eelgrass (*Zostera marina*). *Darker* and *lighter shaded areas* indicate more than 50% and less than 50% coverage of eelgrass, respectively

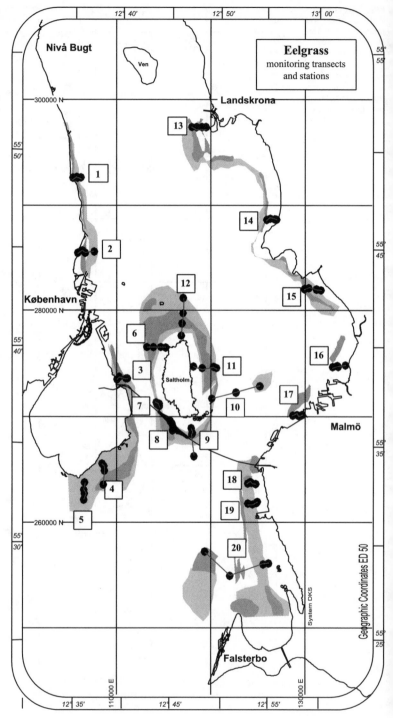

Fig. 7.4. Transects and stations for monitoring of pondweed (*Ruppia* sp.)

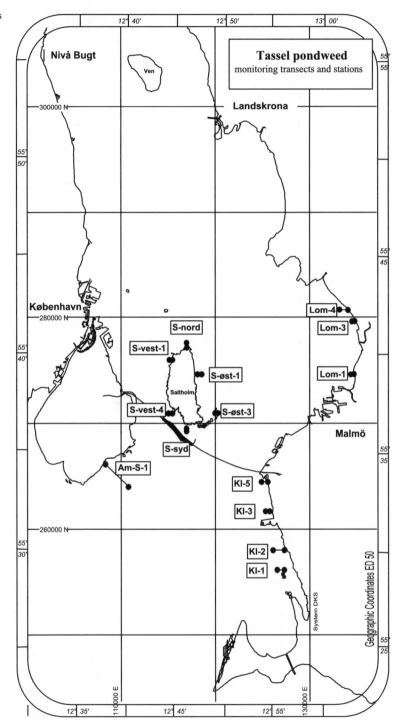

Fig. 7.5. Stations for monitoring of benthic shallow-water fauna

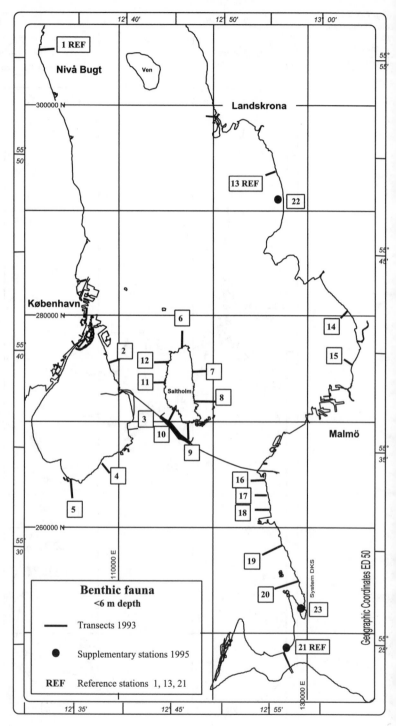

Fig. 7.6. Transects and stations for monitoring of common mussel (*Mytilus edulis*)

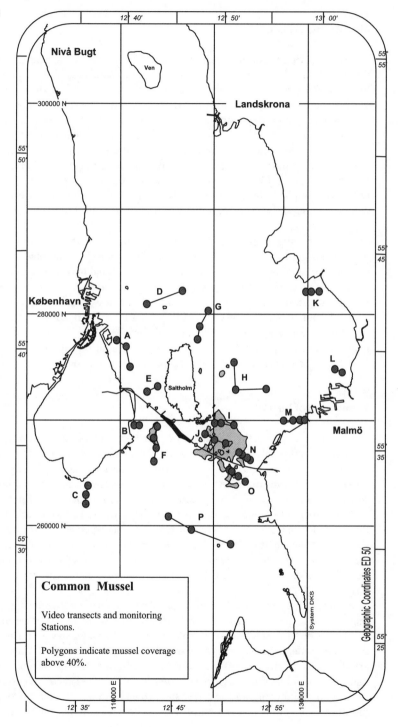

7.2.3
Monitoring of Fish

The construction work must not block the migration of Rügen herring through the Øresund. When this criteria was set, consideration was also made for a number of other fish species, such as eel, garfish and lumpsucker, which also use the Øresund as a migration route. The herring which spawn in the western Baltic use the Øresund as their most important migration route between their summer feeding grounds in the Skagerrak and their spawning ground. The monitoring carried out has shown that the herring move into the Øresund from the Kattegat during the early autumn (August–September), with large parts of the breeding stock spending the winter there, mainly in the deeper areas around Ven. During the spring months there is rapid migration from the Øresund (completed in March) towards the spawning grounds in the western Baltic. After spawning the herring migrate back to the Skagerrak, but the Øresund is not their only route back.

During the migration seasons of 1995/1996 and 1996/1997 a monitoring cruise was undertaken in the autumn which showed that the herring had migrated into the Øresund and were present in concentrations which corresponded to what had been seen during the baseline studies in 1993–1995. Monitoring expeditions during the spring showed that the herring had left the Øresund for the spawning grounds as expected.

In addition to monitoring cruises this programme also includes continuous logging of the passage of shoals of herring through Drogden. The equipment for this was adjusted during the spring of 1996 and has been in operation since the beginning of June 1996. Several passages of shoals of herring were recorded during this period.

7.2.4
Monitoring of Birds

A reduction of 15% in the population of breeding eiders (*Somateria mollissima*) on Saltholm is acceptable, but no later than 5 years after seabed work is completed the number of breeding pairs must be at least 90% of the number of breeding pairs counted during the baseline studies. As far as other breeding waterfowl on Saltholm are concerned, there must not be a significant reduction in numbers. A temporary reduction in the number of foraging and resting migratory birds on Saltholm as a consequence of the construction work can be accepted. However, it is expected that the number of birds will be re-established no more than 2 years after the completion of the construction work. As far as greylag geese (*Anser anser*) and mute swans (*Cygnus olor*) are concerned, it is accepted that there is a risk of a permanent decline in the number of moulting birds.

As part of the programme, studies have been made of overwintering, breeding and moulting birds, partly at Saltholm and partly on the Swedish coast between Landskrona and Falsterbo. Owing to the hard winter of 1995/1996 and heavy ice formation, including inland waters, the numbers of diving ducks in the Øresund in January and February were large, particularly at the future bridge abutment at Lernacken, where approximately 15 000 tufted ducks (*Aythya fuligula*) were recorded in early February 1996. The intensity of monitoring was therefore stepped up, with weekly counts being done from the air instead of the normal monthly counts. Effects from the construction work were not found. The studies on the Swedish coast were done as preliminary studies, as no extensive bird studies had previously been made on the Swedish side of the Øresund. The preliminary studies showed that construction of the bridge abutment at Lernacken may have an impact during periods with ice.

7.2.5
Monitoring of Beaches and Coasts

7.2.5.1
Beaches and Bathing Water Quality

The quality of the beaches and bathing water along the coasts in the Øresund outside the inner impact zone must not be changed to such an extent that the quality requirements for bathing water cannot be complied with. By agreement with the Municipalities of Copenhagen and Dragør measurements of *Escherichia coli* and measurements for suspended matter are being made in addition to existing monitoring of bathing water along Amager Beach and along Dragør's north and south beaches.

Øresundskonsortiet has displayed information on the construction work in progress and its possible effects on the beaches in question in co-operation with the local authorities.

The results of the studies carried out on the Danish side in the period up to the end of July 1996 did not show any impact on bathing water quality along the stretches of coast investigated. The bathing water limits for both Secchi depth and *E. coli* were therefore complied with throughout the period. Nor were any infringements of the set limit values recorded for the bathing beaches investigated on the Swedish side.

7.2.5.2
Coastal Morphology

The effects of the dredging work must be limited to the immediate vicinity of the alignment. The work must not result in sanding-up which causes the artificial island to become joined with Saltholm. Only insignificant changes in bottom and coastline conditions on the other coasts are accepted in the direction of more vegetated coast.

The control and monitoring programme for the coast includes the coastlines at Amager and Saltholm on the Danish side and the area around Lernacken on the Swedish side. The programme is carried out every year in June/July and covers a large number of profiles. To this can be added charting by means of aerial photography of the coastline.

7.2.6
Environmental Information System

Requirements were set in connection with approval of the control and monitoring programme by the Danish and Swedish authorities to the effect that the authorities should have on-line access to all relevant environmental data which are gathered in the context of the project. To this end Øresundskonsortiet has developed an environmental information system, Eagle, which underwent testing by the authorities in the spring of

1996 and was put into normal operation in mid-June 1996. Users have on line access to all the data on the system.

Using Eagle it is possible to present the entered data in the form of tables and graphs, as well as on area maps and in picture format. Eagle is based on a geographical information system, and all the data are stored in the system and linked to a set of geographical co-ordinates so that, for example, station positions and excavation areas can be shown on a map of the Øresund.

7.3
Postscript

Since this chapter was presented during the Symposium on Large-Scale Constructions in Coastal Environments in April 1997, the authorities' monitoring programme has been revised (number and position of stations, sampling strategies etc.) for the second year of monitoring. The revision was made with regard to improving the statistical analysis of possible effects caused by the construction work. The revision was based on the experience from the first year of monitoring, July 1996–June 1997, and recommendations of the International Advisory Expert Panel for the Øresund Fixed Link.

7.4
List of Words and Expressions

- *Control area:* The area more than 7.5 km north and south of the link alignment.
- *Drogden:* The navigation channel on the Danish side of the Link alignment.
- *Inner impact area:* The area stretching 500 m either side north and south of the link alignment.
- *KSÖ:* The Control and Steering Group of the Øresund Fixed Link.
- *Lernacken:* The bridge abutment on the Swedish side.
- *Outer impact area:* The area 7 km north and south of the inner impact area.
- *Secchi depth:* A definition for the vertical visibility in the water, measured by a white disc (Secchi disc).

References

Miljø- og Energiministeriet, Trafikministeriet, Kontroll- och Styrgruppen för Øresundsförbindelsen (1996a) Semi-annual report on the environment and the Øresund Fixed Link's Coast to coast installation spring 199--31 Dec. 1995. (Halvårsrapport om miljøet og Øresundsforbindelsens kyst-til-kyst anlæg. Halvårsrapport om miljön och den fasta förbindelsen över Øresund), pp 44

Miljø- og Energiministeriet, Trafikministeriet, Kontroll- och Styrgruppen för Øresundsförbindelsen (1996b) Second semi-annual report on the environment and the Øresund Fixed Link's coast to coast installation January–June 1996. (2. halvårsrapport om miljøet og Øresundsforbindelsens kyst-til-kyst anlæg. 2:a halvårsrapporten om miljön och den fasta förbindelsen över Øresund), pp 43

Miljø- og Energiministeriet, Trafikministeriet, Kontroll- och Styrgruppen för Øresundsförbindelsen (1997a) Third semi-annual report on the environment and the Øresund Fixed Link's coast to coast installation July–December 1996, pp 22

Miljø- og Energiministeriet, Trafikministeriet, Kontroll- och Styrgruppen för Øresundsförbindelsen (1997b) Fourth semi-annual report on the environment and the Øresund Fixed Link's coast to coast installation January–June 1997, pp 22

Folketinget (1991) Act about the construction of a Fixed Link across Øresund. (Lov om anlæg af fast forbindelse over Øresund). Act no. 590, 19 August 1991

Governmental decision according to the Water Act (1994) (Regeringsbeslut om tillstånd enligt

vattenlagen, M93/4027/4). 16 June 1996
Governmental decision according to the Natural Resources Act (1994). (Regeringsbeslut om tillstånd enligt lagen om hushållning med naturresurser m.m., M92/2372/7). 16 June 1996
Objectives and criteria and the environmental authorities' requirements for the overall control and monitoring programme for the Øresund Fixed Link coast to coast facility, January 1995 (Målsætninger og kriterier samt Miljømyndighedernes krav til det samlede kontrol- og overvågningsprogram for Øresundsforbindelsens kyst-til-kyst anlæg, Januar 1995.). Trafikministeriet og Miljø- og Energiministeriet
Permission for the construction of the Swedish part of the Øresund Fixed Link. Part judgement of Water Rights Court, 13 June 1995 (Tillstånd till uppförande av den svenska delen av Øresundsförbindelsen. Vattendomstolens deldom av den 13 juli 1995.)
Extraction permit in the alignment zone. Ministry of the Environment and Energy, 17 October 1995 (Indvindingstilladelse til Øresundskonsortiet til råstofindvinding i anlægstraceet til den faste forbindelse over Øresund. Miljø- og Energiministeriet, 17. Oktober 1995.)
SEMAC JV (1997a) The authorities' control and monitoring programme for the Fixed Link across Øresund. Water quality. Status Report 1996. Ministry of Environment and Energy (Denmark), Kontroll- och Styrgruppen för Øresundsförbindelsen (Sweden), pp 42
SEMAC JV (1997b) The authorities' control and monitoring programme for the Fixed Link across Øresund. Benthic vegetation. Status Report 1996. Ministry of Environment and Energy (Denmark), Kontroll- och Styrgruppen för Øresundsförbindelsen (Sweden), pp 95
SEMAC JV (1997c) The authorities' control and monitoring programme for the Fixed Link across Øresund. Deep water fauna. Status Report 1996. Ministry of Environment and Energy (Denmark), Kontroll- och Styrgruppen för Øresundsförbindelsen (Sweden), pp 56
SEMAC JV (1997d) The authorities' control and monitoring programme for the Fixed Link across Øresund. Shallow water fauna. Status Report 1996. (Lavtvandsfauna. Tilstandsrapport 1996). Ministry of Environment and Energy (Denmark), Kontroll- och Styrgruppen för Øresundsförbindelsen (Sweden), pp 110
SEMAC JV (1997e) The authorities' control and monitoring programme for the Fixed Link across Øresund. Common mussels. Status Report 1996. Ministry of Environment and Energy (Denmark), Kontroll- och Styrgruppen för Øresundsförbindelsen (Sweden), pp 95
SEMAC JV (1997f) The authorities' control and monitoring programme for the Fixed Link across Øresund. Coastal morphology. Status Report 1996. Ministry of Environment and Energy (Denmark), Kontroll- och Styrgruppen för Øresundsförbindelsen (Sweden), pp 45 excl. appendices

Part III
Environmental Impact and Planning

Europipe II: Early Consultation and Selection of Environmentally Good Alternatives Pay Off

Steinar Eldøy

8.1
Introduction

The planning of a new Europipe II gas pipeline from Norway to Germany started in late 1995, with three possible starting points: the troll gas terminal at Kollsnes outside Bergen, the Kårstø gas terminal in Rogaland (in the south-western part of Norway), or the Sleipner platform in the Norwegian part of the North Sea (Fig. 8.1). According to Norwegian environmental impact assessment (EIA) regulations both in the Petroleum Act and the Planning and Building Act, an EIA was prepared, comparing the different alternative pipeline solutions. In June 1996 the Norwegian parliament approved a plan for developing a major oil and gas field outside the central part of Norway, the Åsgard field, and decided that the gas should be brought onshore to Kårstø for processing. This also implied that Kårstø would be the starting point for a new Europipe II pipeline from Norway to Germany.

The construction of Europipe I was completed in 1994, and the pipeline was put into operation in October 1995. Europipe I was a major challenge both from an environmental and a technical point of view, with the landfall in Germany crossing the Lower Saxony Wadden Sea National Park being the most difficult part. The environmental challenges of Europipe II are quite different from those of Europipe I in Germany, but still there are similarities, and experience from Europipe I has been useful during the planning process. One of the lessons learned from the Europipe I project was that the EIA process should be initiated at an early stage, and should be fully integrated in the project development. This was implemented in the Europipe II project, and, as demonstrated below, this approach has contributed to finding a solution which both minimises conflicts with environmental and fishery interests and at the same time is an economically good solution.

8.2
Alternative Pipeline Routes

For Kårstø as a starting point for Europipe II, three different pipeline routes were evaluated from a technical, economic and environmental point of view (Fig. 8.2). The main alternative was to cross several islands in the municipality of Bokn, the largest called Vestre Bokn, and then continue southwards, west of a group of several small islands called Kvitsøy. The second alternative considered was to lay the pipeline parallel to the existing pipeline corridor from Kårstø to Kalstø at the island of Karmøy, and then continue southwards, west of Karmøy. This alternative would imply using existing pipeline tunnels for fjord crossings. The third alternative considered was to lay the pipeline in

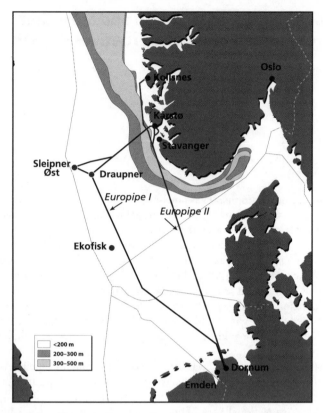

Fig. 8.1. Europipe II from Norway to Germany, showing the different alternative starting points originally considered (Kollsnes, Kårstø, Sleipner)

the sea all the way from Kårstø to Germany. However, with this alternative a quite rough sea-bottom in Boknafjorden and depths down to 580 m would have to be overcome.

Both from a technical and economic point of view, the main alternative via Vestre Bokn would be the preferred alternative. The total cost of this was estimated in the concept phase at 6800 million NOK. The cost of the alternative via Kalstø was estimated at 7100 million NOK, and the alternative in Boknafjorden directly from Kårstø was estimated at 7200 million NOK. Technically it would be a major challenge to place an additional pipeline in the existing tunnels under the fjords between Kårstø and Kalstø, as well as laying the pipeline on the rough bottom with depths down to 580 m in Boknafjorden. However, from an environmental point of view and with respect to fishery interests major challenges where identified with the main alternative via Vestre Bokn.

8.3
Fishery Interests

Just a few years earlier Statoil applied for a permit to construct a condensate pipeline from the Sleipner field to Kårstø along more or less the same route via Vestre Bokn to

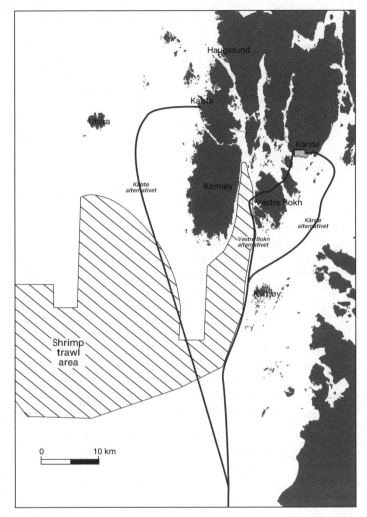

Fig. 8.2. Alternative routes considered from Kårstø. *Hatched area* indicates extension of a shrimp trawling area south and southwest of Karmøy

Kårstø. This was, however, rejected by the government with reference to major fishery interests (shrimp trawling) between Vestre Bokn and Karmøy (Karmsundet) and south-west of Karmøy (Fig. 8.2). With reference to this experience it was decided to establish contact with the fishery authorities and fishery organisations at an early stage. This was done as an initial stage of the EIA process. The aim was both to inform about the project, and to discuss and try to identify possible solutions which could be acceptable also for the fishery interests.

From the outset strong opposition was expressed by fishery organisations and local fishery authorities relating to the crossing of the shrimp trawling areas, but the early dialogue was appreciated. Studies were initiated as a part of the EIA process and in consultation with the fishery interests, in order to clarify details about the shrimp trawling activities, and to identify possible mitigating measures.

Between 20 and 25 shrimp trawlers operate in the area. These are smaller vessels (40–50 ft) with relatively limited motor power. Most of the vessels are only certified to operate within the 12-mile-zone along the coast. Particularly in the area between Vestre Bokn and Karmøy (Karmsundet) the bottom is very soft, and experienced shrimp trawlers judged crossing of a 42-in. pipeline with a shrimp trawl in this area to be very difficult, if at all possible. Trawl tests in the North Sea have documented that trawling across large diameter pipelines does not cause significant problems for the trawling activities. According to these tests (Valdemarsen 1989, 1993), crossing at angels larger than 30 degrees can be done more or less without any impact on the trawl. Crossing at narrower angels can give a temporary tilting of the trawl boards and deformation of the trawl, but the trawl boards can be raised after a few minutes. No hooking and no exceptional damage of the trawl gear was observed during the tests.

Between Vestre Bokn and Karmøy (Karmsundet) the direction of the trawling is north–south. This means that the trawling is more or less parallel to the planned pipeline. As the trawl tests in the North Sea were made with larger and stronger vessels compared to those operating in the shrimp fields around the southern part of Karmøy, it was difficult to insist that the conclusions from the trawl tests could be directly applicable. According to the trawl tests in the North Sea, there could anyway be a potential for tilting of trawl boards and disturbance of the trawl, and because of the narrow area in Karmsundet and the small size of the vessels operating there, other problems could not be excluded.

Statoil therefore decided to indicate a willingness to trench the pipeline in Karmsundet from the landfall at Vestre Bokn towards Kvitsøy, and in the EIA (Statoil 1996) this was stated as a mitigating measure, if the alternative route from Kårstø via Vestre Bokn should be accepted. The fishery authorities and organisations did accept this alternative route, which, as mentioned already, had been rejected just a few years before for the Sleipner condensate pipeline. Later on, more detailed procedures for pipelaying in Karmsundet were agreed with the fishery authorities, starting with the pipelaying and then deciding on the necessity of trenching, backfilling and eventually additional measures based on the results from simple trawl tests.

8.4
Environmental Interests

For the alternative route from Kårstø via Vestre Bokn, two landfall solutions were originally considered: one at the western side of Vestre Bokn (Trosnavåg) and the other near the south-western end of the island. The southern solution was primarily considered as a potential solution for reducing conflict with the shrimp trawling between Vestre Bokn and Karmøy. However, the EIA showed that this route would cross areas registered as being of local and regional value with respect to landscape and nature conservation values, outdoor recreation and some registered cultural heritage sites. Also, from a technical point of view this solution would be quite challenging. As a satisfactory solution had been found with respect to the fishery interests, there was actually no need to consider this solution further, and it was therefore chosen to go for the northern alternative (Trosnavåg) as the landfall site at Vestre Bokn.

The onshore route from the Kårstø terminal via Vestre Bokn, as considered in the EIA, would cross a small island called Ognøy. This northernmost island in the municipality of Bokn was until a few years ago set aside for industrial purposes. However, no industry was established on the island, and during the EIA process it became clear that Ognøy was an area with a number of different conservation interests. Ognøy is registered by the regional nature conservation authorities (the County Governor of Rogaland) as being among the 14 cultural landscape areas of highest conservation value in the county (Fig. 8.3). The area is considered as a well-kept oceanic heather landscape of

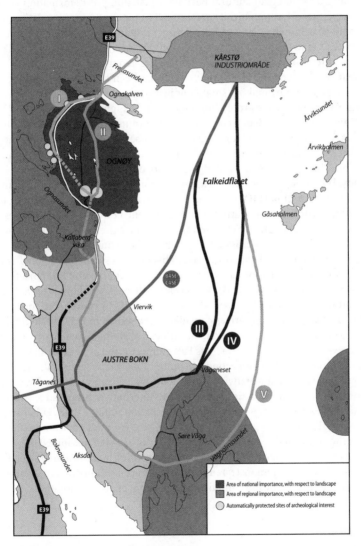

Fig. 8.3. Alternative routes considered for crossing Ognøy or Falkeidflæet from Kårstø, and valuable landscapes and cultural heritage sites along the different routes. The route proposed by Statoil across Falkeidflæet is given as base case

national conservation value. Oceanic heather was previously a common landscape type along the western coast of Norway. However, because of reduced sheep grazing and reduced burning of the heather, it is now regarded as a threatened landscape type. In the EIA it was concluded that the landscape would probably be significantly influenced by the pipelaying, unless mitigating measures could be successfully implemented. Also other vegetation types of high conservation value are found on Ognøy. Such areas would not be directly affected by the pipeline route for Europipe II, but generally they strengthen the conservation value of Ognøy.

According to Norwegian cultural heritage legislation pre-Reformation (1536), cultural heritage sites are automatically protected. The pipeline route alternative via Vestre Bokn would as originally planned have a direct impact on a number of such cultural heritage sites. Ognøy is particularly rich in cultural heritage sites (Fig. 8.3). The pipeline route considered would cross several sites from the Stone Age (more than 10 000 years old). Along the rest of the route it has been possible to avoid most of the cultural heritage sites by making smaller adjustments.

Because of the landscape protection values and the high number of cultural heritage sites at Ognøy, Statoil decided to consider another route directly from Kårstø to Austre Bokn. This route avoids Ognøy completely, and crosses the sea area east of Ognøy called Falkeidflæet. Technical studies and cost estimates concluded that this could be a feasible solution. The costs were concluded to be more or less equal to the solution of crossing Ognøy. When the Norwegian parliament (Stortinget) considered the Plan for Installation and Operation (PIO) of Europipe II, it was underlined that a route avoiding Ognøy reduced conflicts with nature conservation and cultural heritage interests. Later on, however, the coastal authorities raised an objection to the route crossing Falkeidflæet, arguing that this route will cross an area which they consider as suitable for future establishment of a regional anchoring area. This idea to establish a regional anchoring area has been questioned by other authorities. However, because of this objection Statoil initiated additional studies to identify and evaluate alternative solutions both for crossing Ognøy and for crossing Falkeidflæet. An additional EIA was prepared for these solutions (Statoil 1998). The alternatives considered are shown in Fig. 8.3. No final conclusion has so far been reached. However, the additional studies have clearly documented that the previously selected route across Falkeidflæet is the preferable alternative from a technical, environmental and economic point of view. The Ognøy alternative is still considered to be the most negative solution from an environmental and cultural heritage point of view, and the additional costs of this alternative compared to the proposed route crossing Falkeidflæet will be in the order of 280 million NOK. Other routes further east in Falkeidflæet will also be more negative from an environmental point of view, and the additional costs will be in the order of 110–220 million NOK.

Whatever the final outcome is with respect to routing across Falkeidflæet, the planning of Europipe II has demonstrated that respecting environmental interests does not necessarily lead to technically complicated and more expensive solutions.

8.5
Conclusions

In planning the Europipe II gas pipeline from Kårstø in Norway to Germany, extensive contacts with the fishery authorities and fishery organisations, as well as contacts with environmental and archaeological authorities during the EIA process, resulted in a pipeline route and technical solutions which satisfy both fishery interests and conservation interests.

The modifications, including a rerouteing avoiding Ognøy and trenching in the narrowest parts of the shrimp trawling area, are from an environmental and fishery point of view a considerable improvement.

From a cost point of view, the following conclusions can be made:

- Acceptance of the route via Vestre Bokn (with respect to fishery interests) means a reduction in the costs of 300–400 million NOK.
- Changing the route from Ognøy to Falkeidflæet to avoid landscape protection and cultural heritage interests did not cause any increase in the cost estimates. The Ognøy alternative is actually calculated to be 280 million NOK more expensive.
- Modifying the onshore route to avoid archaeological sites did not influence costs significantly (actually a reduction of ca. 1 million NOK was foreseen).

The planning of Europipe II and the EIA process have led to a pipeline route and a technical solution which is significantly cheaper compared to other alternatives considered. The result is thus a solution which combines environmental goals, technical requirements and good economy. Based on the experiences from the planning of Europipe II, the following general conclusions and recommendations can be made:

- In order to identify potential conflicts with other interests, the EIA should be initiated as early as possible as an integrated part of the project development.
- Early dialogue with affected parties and relevant authorities is important for creating trust and confidence, identifying solutions and achieving compromises.
- Caring for the environment and third party interests do not necessarily increase costs, provided the necessary changes and modifications are made during the engineering phase.
- Integrating the EIA into the project development may contribute to achieving beneficial solutions both from an environmental and an economic point of view.

References

Eldøy S (1999) The Europipe development project: environmental challenges of laying a natural gas pipeline through the Wadden Sea. In: Prentice, R.C. and Jaensch, R.P. (eds.) 1997. Development policies, plans and wetlands. Proc. of Workshop 1 of the International Conference on Wetlands and Development held in Kuala Lumpur, Malaysia, 9–13 Oct. 1995. Weltands International, Kuala Lumpur, pp 53–59

Planungsgruppe Grün (1993) Anlandung durch die Accumer Ee. Rahmenbetriebsplan Juni 1993. Teil D: Umweltverträglichkeitsstudie und Landschaftspflegerischer Begleitplan.

Planungsgruppe Grün (1994) Verlegung einer 2. Rohrleitung durch die Accumer Ee. Vergleich der Eingriffswirkungen

Statoil (1996) Konsekvensutredning Europipe II (Environmental impact assessment Europipe II. In Norwegian)

Statoil (1998) Tilleggskonsekvensutredning – alternative traséer Kårstø-Vestre Bokn (Additional environmental impact assessment – alternative routes Kårstø–Vestre Bokn. In Norwegian)

Valdemarsen JW (1989) Trawling across pipelines. Paper to 1st Int. Conf. on Fisheries and Offshore Petroleum Exploration. Bergen, Norway, 23–25 Oct. 1989

Valdemarsen JW (1993) Tråling over 40" rørledning – virkning på trålredskap. (Trawling across 40-in. pipeline – impact on trawl gear. In Norwegian). Fisken og Havet nr. 11, 1993

CHAPTER 9

Seawater Desalination Plants: Heavy Coastal Industry

Thomas Höpner

9.1
Introduction

The global capacity of desalination plants, which is now in the range of 16 million m^3 day^{-1}, is growing annually by a capacity of about 2 million m^3 day^{-1} (Wangnick 1994). This means an annual increase in about five of the large-type plants which are mainly situated at coasts. Desalination plants provide a basic need to population, industry and agriculture. This seems sometimes to evoke the impression that desalination plants are in all means environmentally compatible. Even in fundamental texts, environmental aspects are often missing (e.g. Al-Gobaisi 1994, 1995; Awerbuch 1994). Papers on environmental impacts are in a minority and most of them come from the USA (Kipps 1991; Herbranson and Suemoto 1993; Del Bene et al. 1994,) and papers showing a corresponding title are rare (e.g. Al-Azzaz et al. 1989; El Din et al. 1994, Mannaa 1994, Mickley 1995). Anyone who looks for the keyword "desalination" in the otherwise very comprehensive *Coastal Zone Management Handbook* (Clark 1996) will be surprised to find merely a very little comment on desalination brine among other (suspected) impacts on seagrass meadows. This is an excellent indicator of the fact that desalination plants are usually left out of consideration. On one hand, the nature of desalination plants does not justify this; on the other, there is no reason that desalination plants must fear coastal management and environmental impact assessments (EIAs) more than power stations or other coastal industrial works.

Because of the water scarcity of the Middle East countries and the rapid development of industry, cities and agriculture along the coast, a rapid increase in water consumption is expected (Awerbuch 1994). Even if strategies of water re-use become more effective, the additional demand cannot be fulfilled by conventional and regenerative water supply (Magid 1995). Replacement of conventional sources and provision of additional demand can be expected only from seawater desalination (Al-Sofi 1993). So a steep increase in the number of coastal desalination plants is expected. No wonder that the "desalination market breaks all records" (Wangnick 1994). In the case of the Arabian Gulf, this exploding development is located around an enclosed sea. This merits special attention. In addition, it is concentrated along the most sensitive coastal zones.

Regarding fuel and power consumption, use of additives, area consumption and other effects, desalination plants are heavy industry with waste production and chemical emissions into the air and the water. Price and Robinson (1993) list desalination plants on a level with other emittents and among "major environmental pressures of the Gulf". The marine environment may be affected in the first instance by the brine discharged, by the additives or their conversion products and by the corrosion products. The emission load and the environmental impact are highly site specific, depending on the type and size of the plant and on the character and conditions of the environment.

Facing a worldwide incomparable accumulation of seawater desalination plants along the west and south coast of the Persian Gulf, the governments and rulers of the Gulf countries are starting to reflect past opinion (Al-Gobaishi 1995) and to start the first, still general, environmental impact pilot studies. One of the very first was conducted by the author's group (Hoepner and Windelberg 1996).

9.2
The Legal Context

In the European industrialised countries and many other countries worldwide, new projects likely to have significant effects on the environment are subject to an assessment procedure. The EC directive of 27 June, 1985

> "... shall apply to the assessment of the environmental effects of those public and private projects which are likely to have significant effects on the environment ... project means (also) the execution of constructive works. The EIA will identify, describe and assess in an appropriate manner the direct and indirect effects of a project on the following factors: – human beings, fauna and flora; – soil, water, air, climate and the landscape; – material assets and cultural heritage".

An annex specifies: "Thermal power stations and other combustion installations with a heat output of 300 megawatts and more..." (EC 1985). This means that in Europe thermal desalination plants are duly subject to EIA. Experience is lacking, but since in Mediterranean Europe a powerful desalination development must be expected, examples of EIA are needed.

In the USA, in 1969, the National Environmental Protection Act (NEPA) laid the groundwork for the establishment of several government institutions such as the Environmental Protection Agency (EPA), the Council on Environmental Quality (CEQ) and the National Oceanic and Atmospheric Administration (NOAA). The EPA and CEQ are relevant because they have specified EIA procedures and issued several guidelines and recommendations. US environmental policy with special attention to EIA procedures is described in detail in the NEPA, Section 101 and 102. Among other things, it is stated there that "it is the continuing responsibility of the Federal Government to use all practicable means to preserve historic, cultural, and natural aspects of our national heritage, to achieve a balance between population and resource use and enhance the quality of renewable resources and approach the maximum attainable recycling of exhaustable resources" (US Government 1970).

Section 102 of the NEPA is the basis of the obligatory application of EIA procedures. In particular, it requires all agencies of the Federal Government to

> "... include in every recommendation or report on major Federal actions significantly affecting the quality of the environment, a detailed statement by the responsible official on (1) the environmental impact, (2) any adverse environmental effects which cannot be avoided, (3) alternatives to the proposed action... The Federal actions obligatory to be analysed under NEPA comprise: projects such as power plants, programs such as coastal zone management programs..."

This means that in the USA EIA is obligatory for (thermal coastal) desalination plants. At present, there is no publication listing the countries with environmental legislation and EIA application worldwide. However, in the Middle East nearly all Gulf Cooperation Countries (GCC) are in the process of adopting or have already adopted corresponding environmental laws and have founded the corresponding institutions. Above all, the GCC play an important role in harmonising protection activities and in encouraging local environmental activities. Most recently, the Middle East Desalination Research Centre (Muscat) was founded. It may be expected that it develops to a core institution of cooperation in the interest of environmentally adapted technologies.

Among other arguments, such as avoiding pollution in advance, the application of EIA serves increasingly as an argument for international financing. Additionally, it helps to establish a good reputation outside the country and is nowadays a selling promoter of environmentally sensitive export products such as food or chemicals.

9.3
Coastal Morphological Impacts

The impacts on coastal morphology can be divided into transient and permanent ones. Transient ones are those bound to the phase of construction. They are not treated here. Among the permanent ones are morphological changes at the sites of the plants, e.g. the effects of the seawater intake and concentrate outfall constructions. Since the hydrological and sedimentological dynamics may be affected, the impacts on the coastal ecosystem may be severe.

A large desalination plant with a capacity of 0.3×10^6 m^3 of freshwater per day takes in 2.4×10^6 m^3 seawater and releases 2.1×10^6 m^3 heated and salt-enriched seawater per day. The water taken in must be as poor as possible in suspended matter which is achieved by wide, deep and long channels through which the water flows very slowly and without turbulence. Since water from some distance from the coast is clearer, the channels extend far into the sea, sometimes for more than 1 km. For security reasons sluice doors are built which can be closed during rough weather and extreme water levels. To avoid the entering of sea animals, plants and floating objects, grids and sieves are installed. Taking these requirements together, an intake is a large-scale building exhibiting a severe influence on the coastal dynamics. As a sign of a general re-thinking, Morton *et al.* (1996) describe the intake construction of Station B of the Ras Abu Fontas Desalination plant (Quatar) where the proper intake is located 2 km off the coast and linked to the Plant by pipelines buried in the sea floor.

In contrast, much less expense is usually spent on the outfall. To save money and to remove the proper outfall as far as possible from the point of intake, natural tidal channels, river mouths, lagoons and mangrove swamps are used to guide the discharged water away from the intake. The purpose is to avoid recirculation effects between the intake and the outlet, which clearly is not so easy when the currents change with the tides and the winds. These aspects caused earlier desalination plants to be located within tidal channel and lagoon systems rather than at the open coast.

9.4
Some Technical Information About Thermal Desalination

This section is rather to help the reader understand the principles of thermal desalination than to describe the technical design of thermal desalination plants. For a more extended discussion see Buros (1990). Over 60% of the world's desalted water is produced with heat to distill freshwater from seawater. For this to be done economically in a desalination plant, the boiling point is controlled by adjusting the atmospheric pressure of the water being boiled. The reduction of the boiling point is important in the desalination process for two major reasons: multiple boiling and scale control. To significantly reduce the amount of energy needed for vaporisation, the distillation desalting process usually uses multiple boiling in successive vessels, each operating at a lower temperature and pressure. The process which accounts for the most desalting capacity is multi-stage flash (MSF) distillation. Seawater is heated in a vessel called the Brine Heater and flows into another vessel, called a Stage, where the ambient pressure is such that the water will immediately boil. The sudden introduction of the heated water into the chamber causes it to boil rapidly, almost exploding or flashing into steam. The steam generated by flashing is converted to fresh water by being condensed on tubes of heat exchangers that run through each stage. The tubes are cooled by the incoming feed water going to the brine heater. This, in turn, warms up the feed water so that the amount of thermal energy needed is reduced. MSF plants have been built commercially since the 1950s. They are generally built in units of about 4000 to 30 000 $m^3 day^{-1}$.

Multiple effect distillation (MED) also takes place in a series of vessels (effects) and uses the principle of reducing the ambient pressure in the various effects. This permits the seawater feed to undergo multiple boiling without supplying additional heat after the first effect. MED plants are typically built in units of 2000 to 10 000 $m^3 day^{-1}$.

The vapor compression (VC) distillation process is generally used for small- and medium-scale units. The heat for evaporating the water comes from the compression of vapor rather than from the direct exchange of heat from steam production in a boiler. VC units are usually built in the 20–2000-$m^3 day^{-1}$ range. They are often used for resorts, industries, and drilling sites where freshwater is not readily available.

9.5
Chemical Impacts

Chemical emissions into the sea are less well known than chemical emissions into the atmosphere which are widely discussed in connection with power plants. They consist a list of components:

- Corrosion products (metals)
- Antiscaling additives (polycarbonic acids, polyphosphates)
- Antifouling additives (mainly chlorine and hypochlorite)
- Halogenated organic compounds formed after chlorine addition
- Antifoaming additives
- Anticorrosion additives
- Oxygen scavngers (sodium sulfite)
- Acid
- The concentrate
- Heat

Additives may be discharged in the original form or as reaction products. Additives may react among each other, an option not examined hitherto. Similarly, synergistic effects on the environment by the materials discharged have not been examined so far.

To understand the importance of these emissions a detailed discussion on chemical nature and environmental toxicities is necessary which goes beyond the limitations of this chapter. One of the discharges (9) should be discussed briefly, however. Apart from additives and reaction products, desalination plants discharge the same load of seawater constituents as taken in. The only difference is the concentration. According to Morton *et al.* (1996) a typical product recovery is 10% of the amount of raw water taken in. Then the salinity of the concentrate is 1.1 times higher than the raw water salinity. This means, for instance, discharge of a concentrate of 44 units of practical salinity (psu) into a seawater of 40 psu (to choose conditions which are typical of the shallow southern coast of the Gulf).

On the one hand, it is widely accepted that a marine biocoenosis tolerates only salinity changes of 1 psu (EPRI 1994, cited in Mickley 1995), and conservative discharge recommendations follow this line (Del Bene *et al.* 1994). On the other hand, in hot and arid zones evaporation produces elevated and changing salinities exceeding this tolerance by far. Salinity increases by evaporation are generally larger the lower the water depths. Local salinity changes with depth and time and depends in addition on solar irradiation, wind, tidal regime, water exchange between shallows and offshore waters and other influences. Hence, the relevance of salinity increased by a concentrate outfall has to be examined with careful regard for the local conditions, and usually it is little or absent.

To summarise the emissions of a thermal desalination plant, all emissions have been applied to a production of 1 000 000 m^3 freshwater (Fig. 9.1). This is not an unrealistic figure. At some places it is reached within 2 or 3 days, and along the Gulf coasts every day about 6 000 000 m^3 of freshwater are produced. The flow of matter scheme is a preliminary model using the data of Wangnick (pers. comm.). It needs further theoretical and empirical research. However, at present it is the only scheme available.

The emissions affect the marine ecosystem as a whole and they affect peculiar subsystems. In some cases (e.g. antifoaming additives) the biological effects are unknown. Metal ions are expected to accumulate in the sediments. As a rule, concentrations are of less importance than loads. The extent of the effects depends largely on the sensitivity of the target environment.

9.6
A Case Study: Copper

Thermal desalination plants discharge copper, nickel, iron, chromium, zinc and other heavy metals depending on the alloys present in the process line. In terms of concentrations, copper and iron are highest. The lowest copper concentration value recorded by Oldfield and Todd (1996) is 0.02 ppm (20 ppb, effluent of the Al Khobar desalination plant). To understand that this is critical requires the comparison with natural background concentrations in seawater. Laane (1992) gives 0.07 ppm. This is an oceanic concentration. Data from the Gulf are not available (and it is questionable whether a "natural background" can be expected anywhere in the Gulf, see Sect. 9.10). Hence, copper concentrations in desalination effluents are 200-fold (and more) higher than natural copper concentrations in oceanic seawater.

Assuming 0.02 ppm (20 ppb) copper in the brine (a very low value), a capacity of 500 000 m^3 product per day (Khobar has 576 000 m^3), and a water conversion of 10%, then 100 kg copper will be discharged with 5 000 000 m^3 brine every day at this site. In

Fig. 9.1. Flow of matter and energy of a thermal desalination plant, applied to a production of 1 000 000 m³ freshwater. Data of atmospheric emissions are given for an oil-fired plant which can be considered to be a "worst case". Emissions of gas-fired plants depend on quality of the gas

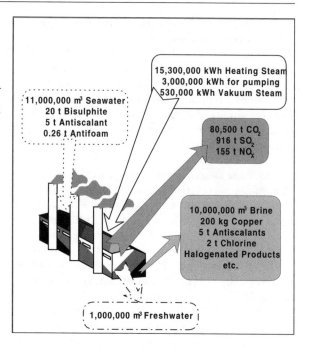

the future, all discharges are applied to 1 000 000 m³ product and 10 000 000 m³ brine discharge, for 0.2 t copper. The heavy metals will adsorb to suspended matter and will sink down causing an accumulation in the sediments. Since the problem is not the concentration but the load, the consequences cannot be mitigated by dilution of the outfall. The load can be distributed over a larger area, but the question of whether a small area heavily polluted is more acceptable than a larger but less polluted area is difficult to answer. International technological efforts try to minimise corrosion (Hassan *et al.* 1994). This is of economical as well as of environmental interest, but nowadays heavy metal discharge is still a main concern.

It must be stated clearly that it is still difficult to connect heavy metal concentrations in seawater and sediments with ecological consequences. The reason is most probably that most of the heavy metals occur in the marine environment in background concentrations (Laane 1992). A second problem is that the metals do not occur in the form of free ions, but in the form of inorganic and organic complexes or they are strongly adsorbed to reactive organic and inorganic surfaces. Commonly, concentrations exceeding the natural backgrounds significantly are considered as environmental pollution even if biological consequences cannot be proven. It is still not possible to set a standard up to which metal pollution is harmless and from which it is harmful.

9.7
A Glance at Reverse Osmosis-Plants

Reverse osmosis (RO) plants are different from thermal plants in some positive and some negative aspects.

1. The overall energy consumption is lower, about one half to one third depending on the technology. Lower energy consumption means less atmospheric emissions.
2. The conversion ratio is higher, about 30 to 35%. This means that the concentration of the brine is higher and attains about 130% of the intake water. At the same time the amount of outlet water is lower than that of a thermal desalination plant.
3. RO plants do not require less chemicals per amount of intake water, but since the conversion rate is higher the emission load compared to produced freshwater is lower. A marked difference is the addition of sulfuric acid to the feed water: 30 mg l^{-1} means about 100 t 1 000 000 m^{-3} freshwater product.
4. The temperature increase is negligible. There are no thermal outfall problems.
5. A true comparison regarding chemical emissions is difficult since data on chemical use are incomplete.
6. Technological problems are less well solved than for thermal plants. This is the main reason for the present predominance of thermal plants.
7. RO plants produce much more solid waste than thermal plants. There are no acceptable solutions for the disposal or re-use of the membrane systems.

Simply stated, thermal plants are favourites of countries rich in fossil energy. RO plants are more common in developed countries. In the Red Sea/Gulf region the number of thermal desalination plants is about 280, the number of RO plants about 112. The ratio of the capacities is 6 800 000 to 425 000 m^3 freshwater per day, respectively (Wangnick, pers. comm.).

9.8
The Targets: Coastal Sub-Ecosystems

The Gulf coasts of the Arabian countries do not vary greatly in geomorphologic structures (Purser and Seibold 1973). Nevertheless, there is a rich diversity of biotopes, some of international nature protection interest (Basson et al. 1981; Sheppard et al. 1992). Sensitive marine Gulf-specific biotopes are, e.g. shallow bays, coral reefs, seaweed meadows, intertidal sand- and mudflats. Sensitive coastal biotopes are mangrove areas, saltmarshes, rocky shores, bird islands, coastal sabkhas, cyanobacterial mats, sand beaches and beachrock flats. The sensitivity of these biotopes to thermal desalination plants is different. To regard the sensitivities properly, a sensitivity index is proposed (see below). According to the variety of biotopes there is also a rich diversity of species (Jones 1986; Price et al. 1993). These properties require site-specific examinations in environmental impact studies.

The use of natural marine resources belongs in a biological context. The Gulf is a productive sea and a rich source of fish and shrimps (Sheppard et al. 1992). Even if fishery is negligible within the gross domestic product, it is present all along the Gulf coast and supplies the food market with goods of high quality and variety. Productive fishery is a good indicator of a healthy marine ecosystem. The number of people engaged in traditional and modern fishery should not be underestimated. Conflicts between the interests of desalination industry and fishery can be expected wherever desalination plants are erected or enlarged.

An impressive and illustrated introduction to the general habitat varieties of the western Gulf has been given by Basson et al. (1981). Jones et al. (no year) demonstrate that even in a sub-area of limited size, the Musallamiya-Dafi Bay system (Saudi Arabia, north of Jubail), a variety of different habitats can be discriminated: terrestrial desert, wetland, rocky, sandy and muddy intertidal zone, shallow subtidal area, open water, coral reefs, islands. An enumeration like this does not point to the narrow mosaic of small and different biotopes which is typical of the well structured coastline stretches. Each of the biotopes has its own properties which make them different in coping with the emissions. In the following text the most typical coastal sub-ecosystems of the Gulf are enumerated, focusing on the flat coasts in the north, west and south.

We have arranged these sub-systems in the sequence from the lowest to the highest sensitivity (Table 9.1). The arguments are coastal ecological ones (arguments like visual sensitivity and others are omitted). The sensitivity index model is based on the coastal sensitivity scale to oil accidents elaborated by Gundlach and Hayes (1978) on the occasion of the *Amoco Cadiz* accident (Bretagne, France) of 1978. Criteria were the sensitivity of the biocoenosis, the danger of oil accumulation and the natural recovery potential. When this model is transferred to the sensitivity to a desalination plant, criteria are sensitivity to the emissions, *but not the ability of the ecosystem to dilute the loads.*

9.9
How to Use the Sensitivity Index

In the case of an EIA for a special project the number of sub-ecosystems is expected to be much less, and the remaining categories will be more similar. It is expected that for a special project several alternative locations will be proposed. Then the sensitivities of these locations will be assessed and compared. The coast of the United Arab Emirates, to take an example, exhibits the categories 1, 6 through 8 and 10 through 15 (see Table 9.1), but sub-categories may be added. The sensitivity profile of a given region is narrower than the profile of a whole sea (like the Arabian Gulf). Normally, a regional sensitivity assessment will be done and two or more sites within one region will be compared. A location at an exposed coastal strip under the influence of a coast-parallel current is less sensitive than a channel-lagoon or shallow bay system. In contrast, a barrier island/lagoon system is a typical case of peculiar sensitivity to desalination plants. The considerable residence time of the water, e.g. within the channels and lagoons, means at the same time that the water discharged stays long within the system. As the salinity gradients show, the influence of the tides is not sufficient to provide effective flushing. It must be expected that pollutants are deposited within the system with significant consequences for the ecosystem.

9.10
Effects on the Regional Ecosystem as a Whole

On a geographical map, the Gulf seems to be an enclosed sea with limited exchange with the ocean. In fact, the exchange is high. A sun-driven current system links the Gulf strongly to the ocean. The Gulf ocean circulation (Sheppard *et al.* 1992) is driven by the evaporation within the Gulf and supported by the prevailing winds. By evaporation, salinity increases in the northern part up to 40 psu. The heavy water sinks down and leaves the Gulf as a bottom current. It is replaced by ocean water of about 37 psu which enters the Gulf as a surface current. The current moves along the Iranian coast while

Table 9.1. Index of sensitivities to desalination plant emissions of coastal marine sub-ecosystems. Note, this it not general but refers to the special situation of the western and southern coasts of the Arabian Gulf. It is not transferable to other marine ecosystems without adaptation to corresponding conditions

Index rank and description	Properties
1. High energy oceanic coast, rocky or sandy, with coast-parallel currents	Rapid water exchange limits salinity and temperature increase. Energy input prevents local accumulations. Provision of oxygen, nutrients and energy provides good conditions of biodegradation
2. Exposed rocky coasts	Good water exchange, even in small niches
3. Mature shoreline	Coast-parallel currents. Accumulation of loads in coastal waters is possible. Sediment mobility limits local accumulations of particle-adsorbed matter
4. Coastal upwelling	The danger of stagnant beach-near water is greater than in category 1. Conditions may change seasonally
5. High energy soft tidal coast	Large intertidal areas and large sediment surfaces susceptible to adsorption and accumulation, but good water exchange and sediment mobility
6. Estuaries and estuary-similar systems	Similar to category 5. Load consequences add to those of terrestrial and anthropogenic loads
7. Low energy sand-, mud- and beachrock-flats	High individual numbers at low species numbers. Loads may accumulate because of adsorption to large surfaces. Limited water exchange
8. Coastal sabkhas	Stress biotope. Exposed to irradiation, wind and dust. Degradations work only during the rare inundation periods
9. Fiords	Semi-enclosed deep water bodies of limited exchange. Danger of thermoclines and oxygen deficits. Shelter and breeding areas of sea animals
10. Shallow low-energy bays and semi-enclosed lagoons	Similar to category 7, but exchange is still lower. Load consequences add to natural stress factors like high and changing salinities, changing water level and solar irradiation
11. Algal (cyanobacterial) mats	Wide intertidal areas at very low beach slopes. High biomass density, but mainly internal cycles of matter. Rather unknown importance for sea organisms. The high position in the index is a question of precaution
12. Seaweed bays and shallows	Shares sensitivity of category 10 and bears additionally sensitivity of seaweed and animals which feed from plants, look for shelter and use seaweed for breeding (e.g. dugongs and turtles)
13. Coral reefs	Basis of a species rich community the species of which have different sensitivities. To range reefs among the sub-systems of highest sensitivity means to regard community members of the highest sensitivity, e.g. fish schools
14. Saltmarsh	Saltmarshes share sensitivity of category 7 and exhibit in addition the sensitivity of macrophytes and of animals which inhabit them. Occasional flooding and precipitates limit degradation periods. Influence of loads adds to stress by salt, dryness, irradiation, dust and grazing
15. Mangal (mangrove flats)	Sensitivity is assumed to be close to category 14, but rapid decline of mangrove areas in the past argues for a still higher sensitivity to many impacts

the bottom current runs mainly along the Arabian coast. The circulation is strong enough to allow an annual exchange of one third of the Gulf's water body.

For the Gulf as a whole, brine discharge is absolutely negligible because the natural evaporation is by several magnitudes higher. This conclusion is also valid regionally since the salinity increases by evaporation are highest within the shallow coastal zones where the brine discharge takes place. The discharges of additives and corrosion products, however, may accumulate within the counterclockwise current. Deposition in the sediments is to be expected downstream of the points of emission. Local examinations must show whether a coastal water strip exists which is not a part of this Gulf-wide circulation and how wide it is. Salinity records are expected to be a suitable means to assess it.

Is it possible that there is a limit for the number and the power of desalination plants within a defined sea area? This question is discussed best with the example of the Arabian Gulf which bears the greatest accumulation of number and power of desalination plants worldwide. Regardless of the sizes, the number of thermal plants is about 277, the number of RO plants 112. The combined power is about 5 800 000 m^3 freshwater per day (data from Wangnick, pers. comm.). Figure 9.2 shows only the sites of plants of a power of more than 50 000 m^3 per day. Even such a sketch illuminates the distribution

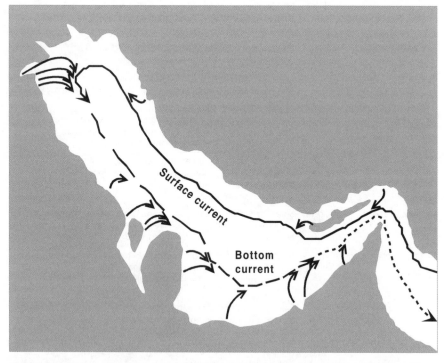

Fig. 9.2. The Gulf area, showing the counterclockwise ring current and approximate situations of the 17 largest desalination plant locations: Bandar Abbas, Queshm, Bushehr (Iran), Doha, Shuaiba, Shuwaik, Az Zour (Kuwait), Jubail, Khobar (Saudi Arabia), Al-Dur, Sitra (Bahrein), Doha, Abu Fontas (Quatar), Mirfa, Umm al Nar, Taweelah, Sharjah (United Arab Emirates)

of the plants only very roughly since the power at one site varies from 50 000 (Bandar Abbas, Iran) to more than 1 000 000 m³ per day (Jubail, Saudi Arabia). The figure shows the counterclockwise ring-current. In the following discussion the question of the number and power of the desalination plants is extended to the combined emissions into this one and the same ring current.

The combined emissions, applied to the power of 1 000 000 m³ freshwater, are compiled in Fig. 9.1. The almost six fold amounts of pollutants are discharged every day into the Gulf or released every day into the atmosphere. Since marine processes are slow, including accumulations and need time to show biological effects, loads calculated per year are more appropriate:

- 440 t copper (which is only one of the corrosion products)
- 11 000 t antiscalants
- 4400 t chlorine

Yearly releases into the atmosphere are:

- 176 000 000 t CO_2
- 2 000 000 t SO_2
- 340 000 t NO_x

The discharges of additives and corrosion products may accumulate within the counterclockwise current. Additive effects in the seawater of the current and depositions in the sediments are to be expected downstream of the points of emission. At present, every desalination project is planned separately from others and without regard for the existing emissions. The author concludes that within a small semi-enclosed sea, this lack of cooperated development cannot be continued into the future.

9.11
Concluding Remarks

With the increasing development of the Mediterranean countries, the wave of desalination plants has now reached Europe. According to the EU directive on the effects of projects on the environment, desalination plants need environmental impact assays. This means that experience in the construction of environmentally more compatible seawater desalination plants is urgently required. Such experience can only be gained by studying foreign examples.

It is not the task of environmental impact studies to prevent the development. The task is rather to adapt projects to the existing limitations and sensitivities. The best way to achieve this is an environmental impact study coinciding with the progress the technical planning and under continuous discussion between project planners and environmentalists.

The number of desalination plants will increase steeply in the near future. Besides economical growth, change of climate will contribute to this increase. In this study it was demonstrated that desalination plants are the origin of many adverse effects on the environment. It is important that planning, construction and operation of desalination plants are accompanied by serious environmental impact analyses. Moreover, it is important that serious and exemplary environmental impact analyses are finished and published within the next few years to serve as models and standards. Otherwise, the number and density of plants will generate not only individual adverse effects, but also cumulative and even synergistic ones – to the disadvantage of large sea areas and semi-enclosed seas.

Acknowledgments. The author thanks Prof. Dr. J. Windelberg (Oldenburg) for elaborating Section 9.2 "The Legal Context", Mr. K. Wangnick (Gnarrenburg) for valuable general and statistical information and Mrs. D. Kremser for the figures. He regrets that conditions of confidence prevent from the use of (unpublished) data and information on distinct desalination plants.

References

Al-Azzaz A, Abrams J, Zaczek S, Awerbuch L (1989) Jeddah stack emissions problems and solutions. Proceedings of the fourth world congress in desalination and water reuse, Kuwait. Vol. I:459–470

Al-Gobaisi DMK (1994) A quarter-century of seawater desalination by large multistage flash plants in Abu Dhabi. Desalination 99:483–508

Al-Gobaisi DMK (1995) The desalination manifesto. Proceedings of the IDA World Congress on Desalination and Water Sciences, Abu Dhabi. Vol. I,3–27

Al-Sofi MAK (1993) Water scarcity – the challenge of the future. Proceedings of the IDA and WRPC World Conference on Desalination and Water Treatment. Yokohama. Vol. II:591–599

Awerbuch L (1994) Perspective and challenges for desalination. Desalination 99:191–194

Basson PW, Burchard JE, Hardy JT, Price ARG (1981) Biotopes of the western Arabian Gulf, 2nd edn. Aramco Dept. of Loss Prevention and Environmental Affairs, Dhahran

Buros OK (1990) The Desalting ABC. IDA (International Desalting Association), P. O. Box 387, Topsfield, Mass. 01983, USA

Clark JR (1996) Coastal zone management handbook. CRC Lewis Publ. Boca Raton, 694 pp

del Bene JV, Jirka G, Largier J (1994) Ocean brine disposal. Desalination 97:365–372

EC (European Communities) (1985) Directive on the assessment of certain public and private projects on the environment. Official Journal of the European Communities L175/40 (July 27, 1985)

El Din AMS, Aziz S, Makkawi B (1994) Electricity and water production in the Emirate of Abu Dhabi and its impact on the environment. Desalination 97:373–388

Gundlach ER, Hayes MO (1978) Vulnerability of coastal environments to oil spill impacts. Marine Technology Society Journal 12:18–27

Hassan A, Al-Sum EA, Aziz S (1994) Storage and retrieval of corrosion data of desalination plant owners. Desalination 97:131–146

Herbranson L, Suemoto SH (1993) Desalination research – current needs and approaches – a US perspective. Proceedings of the IDA and WRPC world conference on desalination and water treatment, Yokohama. Vol. I:209–216

Höpner Th (1996) Long-term observations on natural remediation processes following the 1991 Gulf War oil spill. In: Krupp F, Fleming R (eds.) Establishment of a marine habitat and wildlife sanctuary for the Gulf region. Final Report for Phase III. European Commission (Brussels), National Commission for Wildlife Conservation and Development (Riyadh), pp 127–129

Hoepner Th, Windelberg J (1996) Elements of environmental impact studies on coastal desalination plants. Desalination 108:11–18

Jones DA (1986) A field guide to the sea shores of Kuwait and the Arabian Gulf. University of Kuwait (distributed by Blandford Press, Poole, Dorset, U.K.)

Jones DA, Fleming RM, At-Tayyeb HH (without year). Habitats of the Jubail Marine Wildlife Sanctuary. Senckenbergische Naturforschende Gesellschaft, Frankfurt (Germany) and National Commission for Wildlife Conservation and Development, Riyadh. ISBN 3-929907-21-6

Kipps JA (1991)Seawater desalination along California's central coast. IDA world conference on desalination and water reuse, Washington. Vol. I:1–9

Laane RWPM (ed.) (1992) Background concentrations of natural compounds in rivers, seawater, atmosphere and mussels. Ministry of Transport, Public Works and Water Management, Directorate General of Public Works and Water Management. The Hague (Netherlands)

Magid AMHA (1995) Strategy for the water conservation in the UAE. Proceedings of the IDA World Congress on Desalination and Water Sciences, Abu Dhabi. Vol. VII:452–472

Mannaa AJI (1994) Environmental impact of dual purpose plants. Desalination and Water Reuse 4/1:46–49

Mickley M (1995) Environmental considerations for the disposal of desalination concentrates. Proceedings of the IDA World Congress on desalination and water sciences, Abu Dhabi 1995, Vol. VII:351–363

Morton AJ, Callister IK, Wade NM (1996) Environmental impacts of seawater distillation and reverse osmosis processes. Desalination 108:1–10

Oldfield JW, Todd B (1996) Environmental aspects of corrosion in MSF and RO desalination plants. Desalination 108:27–36

Price ARG, Robinson JH (eds.) (1993) The 1991 Gulf War: coastal and marine environmental consequences. Marine Pollution Bulletin, Vol. 27 (Special Issue). Pergamon Press, Oxford

Price ARG, Sheppard CRC, Roberts CM (1993) The Gulf: its biological setting. Marine Pollution Bulletin 27:9–15

Purser BH, Seibold E (1973) The principal environmental factors influencing Holocene sedimentation and diagenesis in the Persian Gulf. In: Purser BH (ed.) The Persian Gulf. Springer Verlag, Berlin, pp 1–9

Sheppard C, Price A, Roberts C (1992) Marine ecology of the Arabian region. Academic Press, London

US Government (1970) NEPA, US status at large, Vol. 83, 1969. Governmental Press Washington 1970

Wangnick K (1994) Desalination market breaks all records. Desalination and Water Reuse 4/3:25–31

Fixed Link Projects in Denmark and Ecological Monitoring of the Øresund Fixed Link

Henning Karup

10.1
Introduction

Denmark consists of a large number of islands (Zealand and Funen are the two largest) and the mainland Jutland, which is the only part connected with the European continent. Fixed links between the various regions have been discussed during most of the twentieth century. In the 1930s the first bridge from Jutland to Funen and the bridge from Zealand to Falster Island were built. At the same time the first plans to construct a Great Belt link between Funen and Zealand and a fixed link from Denmark to Sweden were presented and, in the following years, new plans appeared in the public debate with regular intervals. In the 1970s the capacity of the bridges erected 40 years earlier became inadequate and new bridges were constructed from Jutland to Funen, and Sealand to Falster.

10.2
Great Belt Link

After a long political discussion and public debate the construction of the Great Belt Link started in 1989. The scheme consists of a four-lane motorway and a twin-track electrified railway. From Sealand to the small island of Sprogø in the middle of the Great Belt the railway is led through a tunnel and the motorway lies on a high bridge. From Sprogø Island to Funen both the railway and the motorway continue on a low bridge. In total, the Great Belt Link is 18 km from coast to coast. The link opened for trains 1 July 1997 and the motorway will be opened in June 1998.

In connection with the public debate on the Great Belt Link project, environmental queries were raised for the first time regarding a construction scheme of this type. The Great Belt Link is erected across the main route of water exchange with the Baltic Sea. It was therefore decided to adopt a so-called zero solution, meaning that the inflow and outflow of water to the Baltic Sea was not to be affected. The zero solution has been made by optimising the link design, and by the use of compensation dredging in the Great Belt.

A comprehensive hydrographic investigation programme was establised together with the control and monitoring programmes dealing with environmental impact on the area concerned. The programmes are conducted by the Great Belt Consortium.

10.3
The Øresund Link

In 1991 the Danish and Swedish governments decided to construct a fixed link from Denmark to Sweden, and the construction of the coast to coast facility started in 1995.

The link extends 430 m from the Danish coast at the airport of Copenhagen where an artificial peninsula has been establised. From there the link will continue under the Drogden navigation channel in an immersed tunnel with a length of 3510 m between the embankments. The immersed tunnel will emerge on a 4055-m-long artificial island south of the existing island of Saltholm. From the artificial island the link continues on a 7845-m-long two-level bridge across the Flinte and Trindel fairways and joins the Swedish mainland at Lernacken south of Malmø. The fixed link comprises a twin-track electrified railway and a four-lane motorway. Denmark and Sweden each on their part undertake to construct the necessary access facilities for railway and road traffic from the fixed link to the existing railway and road networks. The construction work is expected to be finalized in the year 2000.

On the basis of experience gained from the Great Belt Link, a heavy discussion of environmental questions took place during the decision-making process for the Øresund Link. Notably, the spill from the dredging activities in the Great Belt had had some impacts on the local environment. The main environmental questions in connection with the Øresund Link were therefore to balance the impact on the Baltic Sea caused by the blocking effect of the construction, and the impact of compensation dredging activities on the local environment including the shallow waters around the island of Saltholm which is an EU bird protection area. A zero solution for the Baltic Sea would necessitate an increase in dredging in the proximity of the link with a disturbance of the local environment as a consequence.

In order to obtain a balance between the Baltic Sea and the local environment, it is stipulated that the completed structure and the construction works must satisfy a number of environmental objectives and criteria imposed by the Danish and Swedish environmental authorities, and for all the dredging operations, spillage limits in time and space have been set up. In order to ensure that the spillage limits are not exceeded and the quality objectives not violated the authorities, Øresundskonsortiet and the contractors, have developed comprehensive control and monitoring programmes. The aim of the monitoring programme and the results will be described in further details below (see Sect. 10.5).

10.4
The Fehmarn Belt Link

As a part of the Danish-Swedish agreement on the Øresund Link, the Danish government accepted working for a fixed link across the Fehmarn Belt between Denmark and Germany. The experience from the Great Belt Link and the Øresund Link shows an increasing political and public concern for the environmental impact of fixed links on the coastal and marine environment. Seven different technical solution models have therefore been identified for the Fehmarn Belt Link, and comprehensive environmental investigation programmes have been started. The first phase of the investigations was finalized in August 1996 and it has been decided to continue the next phase of the investigations with five solutions models. The second phase of the investigations is expected to be finalized by the end of 1998. The Danish and German governments then have to decide whether they want to proceed with the project. If the two governments agree to do so, an Environmental Impact Assessment procedure will be initiated.

10.5
Ecological Monitoring Programme of the Øresund Link

10.5.1
Objectives

The Danish Public Works Act for the Øresund Link states that the link shall be executed with due considerations of what is ecologically motivated, technically feasible and financially reasonable in order to prevent detrimental effects on the environment. Ecological objectives for the environmental impact should be established together with a control and monitoring programme. The completed structure and the construction work shall satisfy two overall environmental objetives imposed by the authorities (Ministry of Transport and Ministry of Environment and Energy 1995):

- Far field
  - The Øresund Link must not affect the Baltic Sea in such a way that chemical/physical and hence biological changes arise.
- Near field
  - The Øresund Link must only transiently cause conditions in the Øresund that are in conflict with the national plans for the coastal areas. More extensive effects can be accepted in an *inner impact zone* covering the area 500 m either side of the link trajectory measured from the north and south sides of the completed link, respectively, including areas where compensatory dredging is undertaken. Around the island of Saltholm, however, the zone must not extend closer than to a water depth of 1 m. Temporary effects can be accepted in an *outer impact zone* lying 7 km either side of the inner impact zone.
  - The permanent loss of areas as a result of the establishment of the artificial peninsula, artificial island and bridge piles, and permanent effects resulting from local changes in hydrographic conditions can be accepted.
  - Outside the outer impact zone the effects of the construction work must not hinder fulfilment of the objectives and criteria for coastal waters stipulated in the regional environmental plans. As far as the open parts of the Øresund are concerned, the construction work must not reduce the possibilities for establishing an indigenous natural flora and fauna.

10.5.2
Criteria

More extensive effects can be accepted in the inner and outer impact zones, and special criteria for fulfilment of the objectives have therefore been drawn up for a number of selected aspects of nature and the environment (Ministry of Transport and Ministry of Environment and Energy 1995). The first objective implies that the link is to be establised in such a way that there will be no change in the throughflow of water in Øresund nor in the saltwater and oxygen input to the Baltic Sea. This is achieved by optimizing the design of the construction so the blocking effect from the construction is less then 0.5% and then carrying out compensation dredging until unchanged flow conditions have been established. The second objective is achived by careful planning and execu-

tion of the dredging operations in order to keep the spillage percentages below pre-
scribed maximum limits in time and space.

10.5.3
Dredging Operations

The construction work for the Øresund Link involves dredging of approximately
7 million m³ material from the seabed. The overall requirement states that the total
average spillage from dredging operations must not exceed 5% (Ministry of Transport
and Ministry of Environment and Energy 1995). The dredging and reclamation work is
divided into a number of sub-operations (Fig. 10.1). A dredging instruction and a spill
budget for the work have to be approved by the environmental authorities before each
sub-operation starts. The dredging instructions lay down the detailed guidelines for
the extraction work.

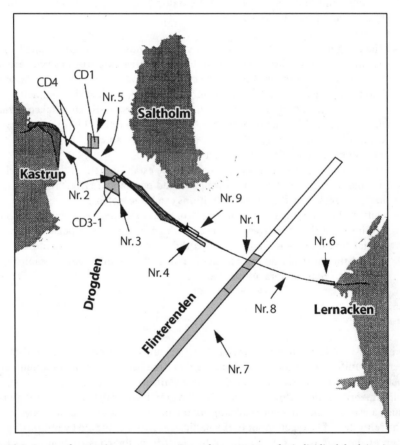

Fig. 10.1. Approved extraction areas as at 31 December 1996. Areas for individiual dredging instructions (Nr. 1–9) are indicated by arrows. *CD1*, *CD3-1* and *CD4* show position of compensation dredging
areas. (Modified from Øresundskonsortiet 1997)

10.5.4
Monitoring Strategy

The Danish and Swedish environmental authorities and Øresundskonsortiet have developed control and monitoring programmes to ensure that the spillage limits are not exceeded and the quality objectives are not violated. The control and monitoring programmes are conducted at three different levels:

- The contractor is contractually obliged to ensure that total spillage limits are not exceeded, and that the requirements for spillage intensity in time and space are fulfilled. The spillage monitoring conducted by the contractor is supervised by Øresundskonsortiet and the environmental authorities in Denmark and Sweden.
- Øresundskonsortiet is responsible for a feedback monitoring programme in order to ensure that timely measures are taken to avoid any risk of violation of any of the authorities' requirements both to the marine environment and to the execution of the work.
- The authorities independently carry out monitoring and control of the environmental impact on water quality, bottom fauna, bottom vegetation, coastal morphology, herring migration, birds and bathing water quality.

The authorities in Denmark and Sweden have to report the results of the ecological control and monitoring programmes semi-anually to the governments.

10.5.5
The Contractors' Spill Monitoring Programme

Spillage measurements are carried out while sailing across the sediment plumes and in profiles in the sediment plumes themselves. In 1996 the measuring was done on a 24-h basis, but during the autumn of 1996 a special method was developed for determining spillage from dredging work done by dredgers equipped with a bucket. The new method determines the spillage percentage using a combination of concrete measurements and model calculations based on the experience gained so far. The aim is to limit measuring activities and so utilize the resources of the measuring vessels more effectively while satisfying the uncertainty requirements with regard to determining the spillage percentage. The new method for determining spillage was put into operation in spring 1997 (Øresundskonsortiet 1996a).

Monitoring of the bed load transport of sediment from dredging operations has been carried out in two stages. The first stage consists of visual observation of the seabed in and around the various dredging areas, and the second stage of more detailed investigation if spilled sediment in thicknesses in excess of 10 cm is found (Øresundskonsortiet 1996b). The non-quality-assured spill data shall within 48 h be stored in an Environmental Information System to which the authorities have direct access. On a weekly basis the contractor has to prepare a report with quality-assured spill data within 3 weeks from the end of the report period.

By the end of 1996 the overall average spillage percentage for all dredging operations was calculated as approx. 234 000 tons or approx. 4.4% of the total dredged amount (approx. 5 300 000 tons (Fig. 10.2); (Danish Ministry of the Environment and Energy, Danish Ministry of Transport, KSÖ 1997a)

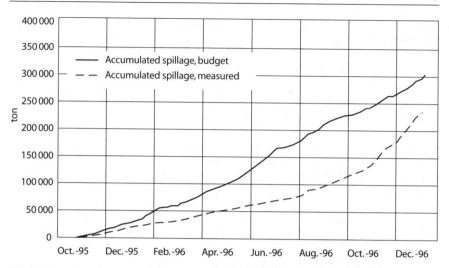

Fig. 10.2. Development over time of measured spillage from dredging and extraction work up to end of December 1996 in relation to budgeted spillage for the period. (Øresundskonsortiet 1997)

10.5.6
Calculation of the Zero Solution

According to the conditions set by the Danish and Swedish authorities, two independent three-dimensional hydrographic models have to be set up for use in calculating the blocking effect of the Øresund Fixed Link and the scope of the compensation dredging which will have to be carried out to reduce the blocking effect to zero. The models were adjusted and tested in the course of 1996 (Øresundskonsortiet 1997). Once final testing of the models is complete it will be possible to start new model calculations. An intensive measuring campaign is planned for 1997 to verify the model. It will be used to assess the effect on currents of building the artificial peninsula and artificial island.

10.5.7
Feedback Monitoring Programme

Øresundskonsortiet uses a feedback programme in connection with the individual dredging operations which involves frequent monitoring of individual, selected parameters in the marine environment around the dredging areas. The feedback programme has the aim of ensuring that early action is taken with regard to the execution of the construction work if there is a risk of the authorities' environmental requirements being exceeded. The programme includes monitoring of sediment, common mussels and eelgrass.

The results from the feedback programme are compared with forecasts made using models which have been updated with detailed information on the individual dredging operations. The models are also used for environmental optimization of the dredging work, with spillage in time and space and a biological evaluation of effects being described and processed. Planning is done with the assistance of a calculation model which includes a hydrodynamic model, a wave model, two dispersion models for calculating spilled sediment and an ecological model for calculating effects on eelgrass, phytoplankton and macroalgae.

Øresundskonsortiet has also set up so-called operational environmental criteria for sediment, eelgrass and common mussels which are stricter and more specific that those of the authorities. This enables corrections to be made for any developments which might otherwise result in an infringement of the authorities' criteria, as well as making it possible to document compliance with those criteria.

10.5.8
Effect Monitoring

As part of their supervision of the construction work the environmental authorities are running a general programme with a view to assessing whether the observed effects of the construction work lie within the frameworks of the expected effects decribed in the Environmental Impact Assessment (Øresundskonsortiet 1995). The programme includes monitoring within the following areas of the Øresund's environment: water quality, benthic vegetation and benthic fauna, fish, birds, bathing water quality and coastal morphology (Ministry of Transport and Ministry of Environment and Energy 1995).

The effect monitoring is long term and is primarily aimed at identifying effects and changes in the environment which are caused by the construction work. In this way the monitoring will form the basis for an assessment of the need for adjustments to the construction work in the longer term. The monitoring also addresses a broader cross-section of the ecosystem than the feedback monitoring, which focuses on fewer variables and the local area around the dredging operations.

10.5.8.1
Water Quality

The monitoring programme is being carried out at four stations in the north, middle and south of the Øresund (Semac 1997a). Each sample is analysed for salinity, temperature and visibility (Secchi) depth, and for concentrations of oxygen, chlorophyll and nutrients, as supporting parameters to help interpret the other sub-programmes. When assessing the oxygen content of the water, oxygen depletion is regarded as beginning at concentrations of less than 2.8 ml $O_2 l^{-1}$ (corresponding to 4 mg $O_2 l^{-1}$). Fish will leave areas with such a low oxygen content. Acute oxygen depletion with values below 1.4 ml $O_2 l^{-1}$ (corresponding to 2 mg $O_2 l^{-1}$) over an extended period can harm benthic fauna so seriously that it will die out (Danish Ministry of the Environment and Energy, Danish Ministry of Transport, KSÖ 1997a).

10.5.8.2
Benthic Flora and Fauna

The programme for benthic fauna includes monitoring of the shallow-water fauna and deep-water fauna together with a separate programme for common mussel. Shallow-water fauna is collected at a small number of transects and stations during the spring, while a more extended sampling takes place in the autumn. The samples are analysed for the occurrence of species, number and biomass. Samples of deep-water fauna are collected annually in April/May at 17 stations with a view to determining biomass, number and size distribution. Common mussel is sampled annually in October/November. Coverage is assessed and biomass, number and size distribution determined. The content of heavy metals in mussels at ten stations is also determined.

The vegetation programme includes monitoring and control of the incidence of tassel pondweed (*Ruppia*), eellgrass (*Zostera*) and sea tangle (*Laminaria*). The programme for tassel pond weed and eelgrass is carried out once a year in August/September and includes aerial photography of the distribution of the vegetation along the coast out to the 6-m-depth contour line. In addition, analyses are made of the distribution and biomass. The monitoring of sea tangle takes place at six localities in August/September using photography and surveying.

10.5.8.3
Fish

The cruises involve surveying the occurrence of herring by means of echo sounding along an extensive system of survey lines throughout the Øresund. The surveys are further supplemented by fishing to establish the occurrence of Rügen herring, and the North Sea herring which has the northern Øresund as the furthest reach of its area of distribution. At Drogden close to the alignment of the fixed link, a stationary sonar system for registering migrating herring shoals was being run in 1996. Echoes were found during the period which presumably represent migrating fish. Proper verification of the echoes with simultaneous fishing is planned.

10.5.8.4
Birds

The bird monitoring programme emphasizes monitoring of the breeding eider population at Saltholm and monitoring of moulting waterfowl (greylag geese and mute swan). More extensive monitoring is performed for occurrence of stageing migrants and wintering waterfowl, mainly tufted duck (National Environmental Research Institute 1994).

10.5.8.5
Bathing Water Quality

A supplementary bathing water programme is being carried out throughout the whole year at the Danish beaches close to the alignment by agreement with the local authorities. On the Swedish side, supplementary bathing water studies were carried out in July and August 1996 by agreement with the municipalities.

10.5.8.6
Coastal Morphology

A total of 65 profiles at right angles to the coast were surveyed on the Danish and Swedish sides in 1996. Thirty of the profiles are located on the coast along Copenhagen, fifteen are located on Saltholm and twenty are on the Swedish side of the Øresund (Semac 1997b). The surveying was done partly with a theodolite from the shore and partly by means of echo sounding from a vessel, with both methods being supplemented by aerial photography. Sediment samples were also taken from twenty four of the profiles for analysis of grain size distribution. Data from earlier studies is being compiled and evaluated in order to compare with the 1996 study to see if any changes of the coastline have occurred as a result of the construction work.

10.6
Discussion and Conclusions

Spillage from the dredging operations is the most important source of impacts on the marine environment during the contruction period of the fixed links. The sediment spill shades vegetation areas in shallow waters and settles on the vegetation and mussels beds. This has potential secondary effects on the feeding resources of breeding eiders and moulting mute swans and greylag geese on the island of Saltholm. The sediment plumes could also potentially prevent migration of Rügen herring through Øresund when the herring during the autumn and winter period migrate from the summer feeding areas in Kattegat/Skagerrak to the spring spawning areas along the German coast in the Baltic Sea.

At the end of December 1996 the Danish and Swedish authorities approved nine dredging instructions and approx. 2.7 million m³ seabed material was dredged with a total spill of 4.4% or approx. 118 000 m³ (Danish Ministry of the Environment and Energy, Danish Ministry of Transport, KSÖ 1997a). The results from the control and monitoring programme for the first year of the construction period show that the dredging was performed without exceeding the spill limits and without violation of the environmental quality objectives (Danish Ministry of the Environment and Energy, Danish Ministry of Transport, KSÖ 1997a). The oxygen content was relatively high in 1996 and there was no risk that the construction works caused or increased acute oxygen depletion (Danish Ministry of the Environment and Energy, Danish Ministry of Transport, KSÖ 1997a).

The concentration of nutrients above and below the pycnocline was generally lower in 1996 than in the previous 10-year period. The contribution of nutrients from the construction works did not lead to a measurable increase in the concentration of nutrients in the water in 1996 (Danish Ministry of the Environment and Energy, Danish Ministry of Transport, KSÖ 1997a). By the end of 1996 the changes in the fauna in the area as a whole did not show a significant reduction in relation to the baseline studies carried out in 1995, and there is no directional trend in relation to the construction work.

The investigation of the benthic vegetation in 1996 showed a decline in the accessible biomass of tassel pondweed in the area south-west and west of Saltholm with the most affected areas in shallow waters less then 2 m and in areas closest to the construction work. The effects in shallow waters indicate that the decline could be caused by the severe ice cover during the winter period 1995–1996 while the effect gradient towards the construction work indicates that the impact is caused by a combination of ice cover and sediment plumes. Provisonal investigations in 1997 show that the tassel pondweed has recovered. The accessible biomass of eelgrass in 1996 was comparable with the level observed in the baseline period of 1993–1995 (National Environmental Research Institute 1997).

The monitoring cruises for the 1995/1996 herring migration season showed concentrations of herring in the Øresund in October which corresponded to the concentrations which were found in the autumn months during the baseline studies in the period 1993–1995, while the monitoring cruises in March and April 1996 showed slightly higher concentrations than during the baseline studies in March and April 1994 and 1995. These higher concentrations in the spring of 1996 point to a later herring migration out of the Øresund owing to the very cold winter of 1995–1996. The concentrations still show, however, that the herring made their usual migration out of the Øresund (Danish Ministry of the Environment and Energy, Danish Ministry of Transport, KSÖ 1997a).

For the 1996/1997 herring migration season two cruises took place in September/ October and November. Another two cruises are planned for the spring of 1997. The first cruise in autumn 1996 showed the concentrations of herring to be slightly different to what was expected in relation to the concentrations found during the baseline studies in 1993–1995 and monitoring in 1995. By the second cruise in the autumn of 1996 the concentrations were back to the expected levels, however.

In 1996, studies of overwintering and moulting mute swans were carried out. The highest number of swans in the moulting period, 1500 individuals, was recorded in early August, which means that the decline observed in the period 1993–1995 continued in 1996 (National Environmental Research Institute 1997a). The decline can presumably be attributed to high mortality in the period January–February 1996, which was probably caused by the harsh winter and ice sheet, and it is highly likely therefore that it was not caused by the construction work. It can be added that the number of moulting swans along the west coast of Scania in July/August was also lower than before, while the number of swans on Saltholm (approx. 2300) in March 1996 was the highest since 1993 (National Environmental Research Institute 1997).

Counts of resting migratory birds on Saltholm in September and October 1996 produced a smaller number of surface-feeding ducks in relation to previous years, presumably because of very dry conditions on the island. The number rose substantially following rain in late October and early November, when the numbers were very similar to those in the preliminary studies. The number of waders showed a similar development (National Environmental Research Institute 1997).

The analyses of the distribution of eiders, greylags and mute swans on Saltholm in 1996 showed that there had been changes in relation to previous years. In all cases there were fewer birds in the immediate area of the alignment and more birds in the areas east, north and north-west of Saltholm. It must therefore be judged very probable that this redistribution is due to disturbances caused by the construction work (Fig. 10.3). As far as all three species are concerned, however, there were still substantial numbers seeking food in and close to the immediate area itself. Subsequent studies showed that the eiders completed their breeding cycle without problems, and that the physical condition of the geese and swans at the end of the moulting period was no worse than in previous years. The birds were therefore able to find sufficient food by using alternative areas further away from the construction work (National Environmental Research Institute 1997).

The overall effects on the birds on Saltholm in 1996 can therefore be characterized as being at the lowest possible level. Thus, it can therefore be concluded that there were

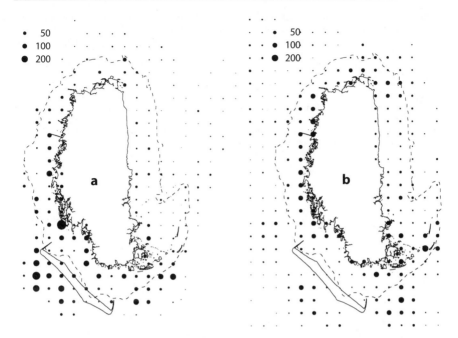

Fig. 10.3. Distribution of eiders around Saltholm on 30 March 1995 (**a**) and 3 April 1996 (**b**) charted on the basis of aerial photography. Charting only shows males. (National Environmental Research Institute 1997)

no infringements of the conditions for the construction works as regards breeding eiders, moulting greylags and mute swans on Saltholm (Danish Ministry of the Environment and Energy, Danish Ministry of Transport, KSÖ 1997).

The results of the bathing water studies carried out have only shown few occasions outside the normal bathing season with cloudy, turbid water with reduced visibility (Secchi) depth at the beach south of the Øresund Link on the Danish coast. No abnormal conditions were observed along the beach north of the Øresund Link and on the Swedish coast.

References

Danish Ministry of the Environment and Energy, Danish Ministry of Transport, KSÖ (1996) Second semi-annual report on the environment and the Øresund Fixed Link's coast-to-coast installation. 1st. half of 1996. 25 pp

Danish Ministry of the Environment and Energy, Danish Ministry of Transport, KSÖ (1997a) Third semi-annual report on the environment and the Øresund Fixed Link's coast-to-coast installation. 2nd half of 1996. 25 pp

Danish Ministry of the Environment and Energy, Danish Ministry of Transport, KSÖ (1997b) Fourth semi-annual report on the environment and the Øresund Fixed Link's coast-to-coast installation. 1st half of 1997

Ministry of Transport and Ministry of Environment and Energy (1995) Objectives and criteria and the environmental authorities' requirements for the overall control and monitoring programme for the Øresund Fixed Link coast-to-coast facility

National Environmental Research Institute (1994) Bird monitoring in relation to the establishment of a
   fixed link across Øresund. Proposal for the Programme. 35 pp
National Environmental Research Institute (1997) Monitoring of moulting mute swans around Saltholm,
   1996. 40 pp
Øresundskonsortiet (1995) Supplementary assessment of the impacts on the marine environment of
   the Øresund Link. 193 pp
Øresundskonsortiet (1996a) Estimation methods for dredged material spill – mechanical dredging.
   63 pp
Øresundskonsortiet (1996b) Bed load monitoring programme status report for dredging instruction
   no. 1–4, October 1995 to September 1996. 63 pp
Øresundskonsortiet (1997) Semi-annual environmental report, July–December 1996. 42 pp (In Danish)
SEMAC JV (1997a) Status report 1996. Water quality
SEMAC JV (1997b) Status report 1996. Coastal monitoring

Towards an Integrated Approach in Environmental Planning: The Europipe Experience

Bastian Schuchardt · Henning Grann

11.1
Introduction

The main task of Integrated Coastal Zone Management is to provide solutions to obtain a balance between ongoing economic development on the one hand, and the integrity of nature and the human environment on the other hand, in other words, a sustainable development of coastal regions. Among other things, this requires a set of specific approaches and methods, an important one being to ensure that environmental protection is integrated into all phases of project planning. Although environmental protection has already become an important part of many engineering and construction schemes worldwide, the development of concepts and experience required to meaningfully and efficiently integrate environmental protection is to date scarce.

A complex methodological approach embodied in an Environmental Protection Plan (EPP) has been developed and implemented for the Europipe project in the coastal zone of northern Germany. Considering the increasing demand for environmental protection worldwide, we contend that the Europipe EPP can in future serve as a tool for similar projects elsewhere.

11.2
The Wadden Sea

The Wadden Sea covers an area of almost 8000 km^2 from Skallingen in Denmark to Den Helder in The Netherlands. It is a unique ecosystem, protected accordingly by several national and international regulations. The flat 5- to 10-km-broad Wadden Sea is characterised by semidiurnal tides ranging in amplitude from 1.5 m to >3.0 m. Extensive intertidal sand and mud flats comprise approx. half the area, the remainder consisting of a network of tidal channels and creeks, salt marshes and barrier islands.

High productivity and biodiversity are amongst the main factors explaining the use of the region as a nursery ground by many commercially important North Sea fish species. In addition, the Wadden Sea is a vital breeding area and a resting area for large numbers of migratory birds from Siberia and Greenland (CWSS 1993). The ecological value of the region is acknowledged worldwide. The national park status of large areas of the Wadden Sea was of major significance to the Europipe project. The region is designated as a wetland of international importance under the Ramsar Convention and as a Special Protection Area under the EC Bird Directive. It is also recognised as a Biosphere Reserve by UNESCO.

11.3
The Europipe Project

In order to satisfy the increasing demand for natural gas in Western Europe a 640 km long pipeline was installed from the Norwegian North Sea to Emden on the north German coast. The landfall of the offshore pipeline took place in 1994 in the German Wadden Sea area. After lengthy consultations with the German state of Lower Saxony which started in 1985, the Norwegian State Oil Company Statoil announced in 1990 that Europipe landfall was planned to take place along the coast of Lower Saxony. During this planning phase it became clear that the pipeline would have to cross the Wadden Sea National Park of Lower Saxony. After evaluating a number of possible landfall routes in terms of a wide variety of technical, safety and environmental criteria, a landfall concept was proposed for the barrier island of Norderney in early 1991. It involved traditional pipelaying methods in a formal Regional Planning Procedure which includes an Environmental Impact Assessment (EIA). A heated public disussion started, fueled by a shift in the political scene in Lower Saxony, whereby attention became increasingly focused on environmental issues. Consequently, several alternative routes were evaluated and discussed in detail (Grann and Schuchardt, in press).

At the end of a very complex planning phase, it was finally decided that the route should cross the National Park of Lower Saxony through the Accumer Ee tidal inlet between the barrier islands Langeoog and Baltrum. In November 1992 Statoil was granted formal planning permission for the scheme, which included the construction of a tunnel below the intertidal flats and dike. After additional efforts had been made to optimise planning with regard to environmental considerations, final permission for construction was issued in October 1993. Work commenced on land in November 1993. In the Wadden Sea proper construction activities took place mainly in 1994 and early 1995, and delivery of gas commenced late in the same year (October 1995).

The pipeline landfall section is about 12 km long, extending from the 3-mile zone seaward of the East Frisian Islands Baltrum and Langeoog, crossing a very dynamic, shallow sandbank area (reef bars), following the Accumer Ee tidal inlet between the islands, continuing into the shallow Accumersieler Balje before reaching the tidal flats. For the purpose of burying the pipeline permanently in the seabed, a trench of varying depth and with a trench bottom width of 12 m was dredged seaward of the tidal flats. In addition, a 100-m-wide vessel access channel had to be cut through the above mentioned sandbank area (see Fig. 11.1). In total some $3.4 \times 10^6 m^3$ of mainly sand was removed to be used for beach nourishment or to be brought to an interim storage area offshore for later backfilling of the trench. Pipelaying into the open trench was performed by a large pipelaying vessel in the tidal inlet and seaward and using a smaller one in the Accumersieler Balje. The large pipelaying vessel could not be used with its normal anchor pattern when operating in the shallow and narrow tidal channel. Instead, 34 anchor piles were pre-installed and used as anchors during pipelaying.

The tidal flats, the foreland and the dike have been crossed using a sub-surface tunnel (length about 2600 m). The tunnel start shaft was located behind the main dike. The physical connection of the offshore pipeline in the trench and the land pipeline in the tunnel was performed in a sub-sea floor tie-in chamber located in the Accumersieler Balje (see Fig. 11.1). For details see Grann (this Vol.).

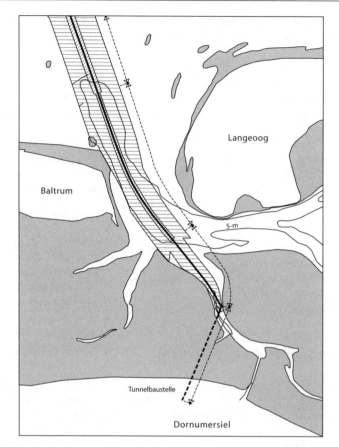

Fig. 11.1. Pipeline route through the National Park in northern Germany with designation of route sections. The "nearshore section" lies north of the island Baltrum outside of map area. The seabed directly affected by dredging measures or anchorages are marked with dashed lines. (Storz and Schuchardt 1995)

11.4
The Environmental Protection Plan

The Wadden Sea National Park was established to protect a wetland of extreme environmental importance and vulnerability. Hence, very stringent environmental requirements have been set by the competent authorities. The Europipe project has received much attention from politicians, authorities, environmental groups, local communities, and the public in general. As a result, environmental issues were identified as being of primary importance for project approval.

Within this setting a detailed and systematic environmental management approach was evidently required to facilitate the smooth running of the project. With the aim of ensuring a high level of environmental performance, Statoil (as operator of a group of companies) implemented an Environmental Protection Plan (EPP) which accompanied every stage of planning, construction and operation. In this way environmental management was introduced into the project, giving environmental expertise a central position within the management team. The EPP (see Fig. 11.2) comprises:

1. An Environmental Protection Programme for the Planning Phase dealing with:
   - Ecological Sensitivity Studies, Environmental Baseline Studies, Environmental Impact Assessments
   - Communication between technical and environmental experts, relation with the authorities, information flow to the public
2. An Environmental Protection Programme for the Construction Phase dealing with:
   - The Ecological Monitoring Programme, Ecological Compensation Measures
   - Environmental Management of Construction Activities
   - Communication between technical and environmental experts, relation with the authorities, information flow to the public
3. An Environmental Protection Programme for the Operation Phase dealing with:
   - The Ecological Monitoring Programme
   - The Environmental Management of Operation, Environmental Audits, Environmental Management Reviews
   - Information and training, relations with the authorities

In the following section we will focus on some special items of the EPP.

11.5
Sensitivity Studies (Route Selection)

Planning of a pipeline route has to take account of the natural, socioeconomic and political characteristics of the region it crosses. At the same time, such a routeing has to comply with construction and safety requirements. Where avoidance of a sensitive area (in this phase defined as protected areas) is not possible, it may be feasible to minimise the impact by adopting special measures. However, in principle at least, engineers will want to avoid areas which present particular construction difficulties whilst environmentalists will want to avoid areas of ecological sensitivity and other groups may focus on other specific items. Demands of this nature can lead to a dilemma in deciding what should take precedence in the final selection. It is obvious that such a decision must ultimately be a political one (Grann and Schuchardt, in press). However, it is essential to involve *all* stakeholders in the decision-making process, which failed partly, in view, in the Europipe project. It becamesobvious during the Europipe project that strategic and conceptual project planning had to take into account all these different concerns and interests in the right balance. Aspects of the environment and of environmental policy were underestimated in the very early planning of the Europipe project due to the lack of qualified information.

The very early analysis of technical, ecological and socio-economic aspects as integrated sensitivity studies offers a promising approach in this respect. Sensitivities may be defined as key components and processes which are critical to the integrity of a system, both of the natural and the socio-economic environment (see De Bie 1996). The early identification of the key components and processes which may be critical to the composition and integrity of both the environment and its function for mankind should be the main aim of such sensitivity studies.

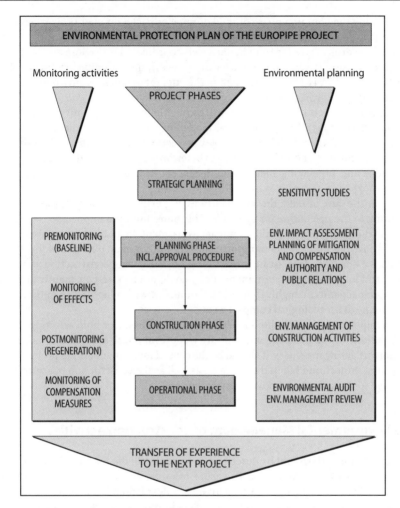

Fig. 11.2. Environmental Protection Plan of the Europipe project

One major conclusion of the Europipe project is that there is a strong need for an improved understanding of the interrelationships between technology, business, politics and the environment and that this understanding has to be introduced in project planning and management. This means that concepts of integrated assessment (see Van Asselt *et al.* 1996) should be further developed and implemented, including open discussions with all parties involved in very early planning stages.

11.6
Environmental Impact Assessment

Since a few years ago Environmental Impact Assessments (EIAs) have become an integral part of approval procedures for large-scale constructions in most European coun-

122 B. Schuchardt · H. Grann

tries. During the planning phase of the Europipe project a number of EIAs have been performed as a result of newly introduced routes or pipelaying methods. Although this phase of the project was far from being a well-coordinated procedure due to the very complex situation (Grann and Schuchardt, in press), the Europipe experience shows clearly that the EIA is a central element in the attempt to enhance environmental protection during all project phases.

Only some aspects of the EIA work will be mentioned in this chapter (see also Storz and Schuchardt 1995). Using numerical modelling to predict hydrological and morphological effects is essential for a complex project such as Europipe. Having the limitations in mind, it can be a powerful tool to determine and evaluate the effects of different routes and pipelaying methods with respect to hydro- and morphodynamics (Bijker 1997).

Due to the lack of data, the EIA had to work partly with worst case assumptions according to the precautionary principle. This underlines the importance of an environmental data base for a qualified impact assessment. In addition, these data should serve as baseline studies for a monitoring programme. A main outcome of the EIA was that on the one hand large-scale impact would occur during construction and on the other hand long-lasting or permanent effects could not be foreseen, mainly due to the kind of impact and the highly dynamic environment with a high regeneration potential of many main ecological compartments.

The importance of EIAs should be underlined, as a basis for both selecting alternative solutions in the planning and authority approval phase and identifying and designing mitigating measures. It should be the central instrument for the project's Environmental Protection Plan in the planning as well as the construction phase and must be linked to an ecological monitoring programme.

11.7
The Environmental Management of Construction Activities

One item of the EPP should be described in some detail: the Environmental Management of Construction Activities. To ensure minimal adverse environmental effects caused by the physical construction activities, a detailed and systematic approach, a methodology which enables identification of environmentally critical work operations, development of effective protection measures and above all development of simple work instructions which translate the sometimes vague environmental issues into concrete action were required.

This can be illustrated by the "zero dumping concept", the general concept of all marine operations in the National Park, meaning zero emission or disposal of any solid waste and liquid effluents in the environment while operating in the area. All sewage and waste water, including rainwater on deck, had to be collected on board, and transported to approved receiving facilities onshore or be brought out of the area. A waste shuttle service was established to handle this. In addition, fuel consumption had to be minimised and only low sulphur fuel was to be used. To meet these (and other) requirements, all vessels (about 90) had to be modified and approved.

For this purpose Statoil used the British Standard BS 7750 "Specification for Environmental Management Systems" as a basis for the development of the required systematic planning approach (though other approaches under EMAS or ISO are also possible). None of the Europipe project contractors was familiar with this environmental planning concept and they were initially all rather sceptical with respect to the practicality of this approach. For this reason Statoil developed a set of guidelines which extracted the essentials from BS 7750 in a form which fairly specifically defined Statoils requirements to the contractors. On the basis of this document, the contractors were requested to develop their environmental plan consisting of the following three levels of documentation:

1. Environmental Management Manual
In this document the contractor states his commitment to a high level of environmental performance and describes how he intends to carry out the environmental management. The manual addresses such items as environmental policy and objectives, organization of environmental activities, personnel qualifications, permits and regulations, environmental audits, reviews etc.

2. Environmental Protection Programme for the Construction Phase
This document provides a considerable amount of details based on the principles and statements given in the environmental management manual. The programme includes identification and description of environmentally critical operations, description of adverse effects of environmentally critical operations, description of protective measures, programme implementation, control and verification, and emergency planning.

3. Operational Control Sheets
These one-page sheets provide simple concrete work instructions for the execution, inspection and control of each of the specific environmental protection measures. They are for the use of the first line supervisors, the inspectors and the various craftsmen actually doing the job. The sheets cover definition and description of the activity, plan for work instruction, instruction for job execution, responsibility/authority, control and verification, and handling of deviations.

The above environmental management approach was very demanding, not only in terms of the resources required to develop the environmental plan and action programmes, but also in terms of training of contractor personnel and gaizing contractor acceptance and commitment. Regular site inspections were carried out by environmental inspectors to ensure that the agreed environmental plans were in fact adhered to. These inspections were mostly done jointly by Statoil and the contractors personnel. To assist in the implementation of the environmental protection measures, a joint committee of Statoil/contractor staff and environmental specialists, the so-called environmental network, was established. This committee met regularly throughout the construction period to make status reports and to facilitate the exchange of experience between contractors.

In conclusion, it can be stated that the considerable environmental management efforts have been well spent. Only a few very minor environmental incidents occurred and a considerable amount of useful experience has been gained for use in future construction projects.

11.8
Ecological Monitoring Programme

An extensive Ecological Monitoring Programme, aimed at the identification and documentation of the main impacts on the environment, was an important part of the EPP of the Europipe project. This programme, required by the approval authority, covered several disciplines including hydrodynamics, morphodynamics, sedimentology and biological aspects such as zoobenthos, fish and birds (for overview see Vollmer 1997). However, the limitations of such a programme should be mentioned: it is and it will be very difficult and partly impossible to identify relatively weak and temporal effects and to distinguish between the natural variability and the project's impact in this highly dynamic environment. This may be illustrated by the number of breeding pairs of the endangered little tern (Frank and Grünkorn 1996) on the eastern tip of Baltrum. A possible impact has been foreseen in the EIA and several mitigation measures have been implemented (such as construction of anchor piles clearly before the colony has been founded and the construction of a fence to stop curious tourists, which might be attracted by the pipe-laying activities). Looking at the results of the Ecological Monitoring Programme, a steep decrease in the number of breeding pairs from 1993 to 1994, the year of construction, can be seen (Frank and Grünkorn 1996; Fig. 11.3). Taking the long-term dynamics into account, the results are less clear: is the decrease due to the construction activities or due to natural variability? However, it is evident that in the year after the construction the number of little terns was close to the long-term average and it is thus probable that no long-term impact has occured, as was foreseen in the EIA (Storz and Schuchardt 1995).

Nevertheless, an Ecological Monitoring Programme is an essential feature of integrated environmental management in order to identify the environmental impact and to evaluate the activities and the mitigation measures.

11.9
Environmental Audit

An Environmental Audit was also part of the Europipe EPP. Audit in the present case means that the actual effects documented in the Ecological Monitoring Programme are

Fig. 11.3. Number of breeding pairs of little tern (*Sterna albifrons*) on the eastern tip of Baltrum. (Data from Frank and Grünkorn 1996)

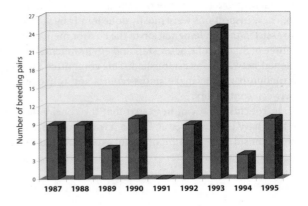

Table 11.1. Dredging volumes in 1000 m³(Bioconsult 1997)

Area	Sediment volume, estimations[a] in EIA Europipe I	Sediment volume, estimations[a] in EIA Europipe II	Sediment volume, real quantities	Difference	Difference (%)
Near coastal	367	375	283	−92	−24.6
Sand bar	1138	1165	1207	+42	+3.6
Tidal channel	1206	1235	1181	−54	−4.4
Balje	213	253	300	+47	+15.7
Total	2924	3028	2971	−57	−1.9
Maintenance dredging	250	250	470	+220	+88.0
Total	3174	3278	3441	+163	+4.9
Substitute fairway	–	–	204	+204	–
Total	3174	3278	3645	+367	+11.1

[a] Including a contribution for unforeseen amounts.

compared with the impingements determined in advance and the reasons for any eventual deviations are analysed. Differences concerning technical data assumed in the stage of the planning and construction period were summarized in an environmental account. As an example, Table 11.1 shows the accounting of the dredged volumes. The aim is to learn for further projects, to improve the methodology of the EIA and Ecological Monitoring Programme and the integration of the different parts of the EPP.

For most of the ecological compartments under consideration the impingements documented in the Ecological Monitoring Programme were similar to those predicted in the EIA (Schuchardt, in press). In some compartments the intensity of these effects was clearly lower than predicted. The three main reasons for this are the necessary worst-case assumptions in the EIA due to lack of data, an overestimation of the sensitivity of some species in the impact area and additional construction activities. An example of the overestimation of the impact is the number of seals. In the EIA it was predicted that all animals would avoid the area for resting during construction. As can be seen from Fig. 11.4, there was a clear reduction in the number of animals but not a complete avoidance of the area. However, some impingements seem to have been underestimated, e.g. the possible increase of the infection rate of black-spot disease in brown shrimps (Knust 1997).

Up until now there have been no indications that the natural ecosystem processes will not recover as predicted in the EIA. However, the results of the ongoing monitoring investigations describing the regeneration process need to be evaluated for a final assessment.

11.10
Conclusions

Since a few years ago, Environmental Impact Assessments (EIA) have become an integral part of regional and other formal planning procedures for large-scale constructions in most European countries. The Europipe experience shows clearly that the EIA

Fig. 11.4. Seal counts in the Accumer Ee impact area (*below*) in comparison with the East-Frisean Wadden Sea area in total (*above*) (Data from National Park Authority, Wilhelmshaven)

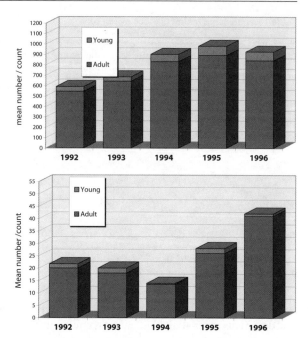

is a central element in the attempt to ensure environmental performance during all project phases. It can be an essential tool to develop measures to reduce the environmental impact. However, it must be integrated into an Environmental Protection Plan, running through all project phases. We have learnt from the Europipe experience that integration of environmental aspects is essential during very early phases of project conception and planning. However, further key issues of "Integrated Planning" are integration of environmental aspects and technical planning, integration of environmental aspects during all project phases, integration of *all* environmental aspects and integration of political, socio-economic, technical and environmental aspects. From the Europipe experience it can be concluded that:

- An EPP is an essential feature of up-to-date project planning. However, a systematic but flexible approach is neccessary.
- Early sensitivity analysis should integrate technical, political, socio-economic and environmental issues.
- An early open discussion with NGOs and the public must be an integral part of project planning.
- A reliable data base is important for all phases of the EPP and might help to avoid worst-case assumptions.
- It is important that environmental expertise is given a central position within the project management team.

Working with a systematic but flexible EPP may ensure high ecological performance at all levels of decision making, project planning, construction and operation. In so doing, an EPP may become an important tool in Integrated Coastal Zone Management, possibly also increasing economic competitiveness for clients, contractors and consultants.

References

BIOCONSULT (1997) Environmental audit of the Europipe project in the National Park Lower Saxony's Wadden Sea. Report compiled by Bioconsult Schuchardt and Scholle for Statoil, Emden, unpublished

CWSS (1993) Quality status report of the North Sea. Subregion 10. The Wadden Sea. Common Wadden Sea Secretariat, Wilhelmshaven

Bijker R (1997) Potentials and pitfalls of hydraulic modelling – does it always support balanced infrastructure developments in the coastal environment? Forschungszentrum Terramare Berichte 1:37

de Bie S (1996) Sensitivity analysis: the approach to integrate ecological and socio-economic impact assessments. In: SPE (Society of Petroleum Engineers) Proceedings of the Int. Conference on Health, Safety and Environment, New Orleans, Louisiana, pp 289–296

Frank D, Grünkorn T (1996) Ökologische Begleituntersuchungen zum Projekt Europipe – Teilprojekt Avifauna – Abschlußbericht 1994 Band I und II. Nationalparkverwaltung Niedersächsisches Wattenmeer Wilhelmshaven, unpublished

Grann H (this Vol.) The Europipe landfall project

Grann H, Schuchardt B (1998, in press) Europipe – approval procedure and Environmental Protection Plan. J. Coastal Res.

Knust R, Ulleweit J (1996) Ökologische Begleituntersuchungen zum Projekt Europipe – Teilprojekt Fische und Krebse – Endbericht. Alfred-Wegener-Institut für Polar- und Meeresforschung Bremerhaven und Universität Bremen, unpublished

Schuchardt B (1998, in press) Environmental audit of Europipe construction in the Wadden Sea. The assumptions of the Environmental Impact Assessment and the results of the Ecological Monitoring Programme – a comparison. J. Coastal Res.

Storz G, Schuchardt B (1995) Verlegung der Gasleitung Europipe im Nationalpark Niedersächsisches Wattenmeer: Ergebnisse der Umweltverträglichkeitsstudie. Wasserwirtschaft 85 (1995), 5:244–248

van Asselt MBA, Beusen AHW, Hilderink HBM (1996) Uncertainty in integrated assessment: a social scientific perspective. Environmental Modeling and Assessment 1:71–90

Vollmer M (1997) The importance of Ecological Monitoring Programmes to large-scale constructions. Forschungszentrum Terramare Berichte 1:38

Part IV
Recourse and Coastal Management

Managing the Ecology and Economy of Modified Estuaries: The Delta Project in The Netherlands

Aad C. Smaal · Alle van der Hoek

12.1
Introduction

The Delta Project

Like most delta-estuarine environments, in its natural state the Dutch Delta region, in which the Rhine, Meuse and Scheldt have their estuaries, contains various complicated ecosystems. These reflect the complex hydrodynamic regime, with fast-flowing water masses in tidal channels, changing estuarine configurations, inhomogeneous tidal and subtidal sediments and salt marshes that flood periodically (Fig. 12.1). The history of

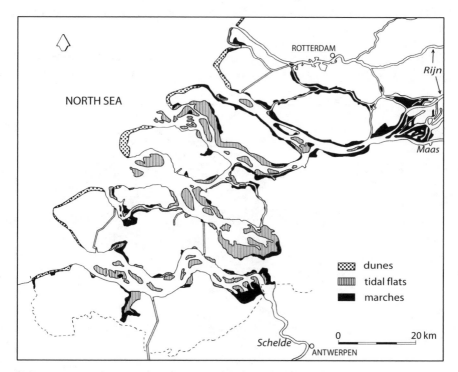

Fig. 12.1. Delta area of the rivers Rhine (*Rijn*), Meuse (*Maas*) and Scheldt (*Schelde*) in the south-west Netherlands in 1950 with salt- and freshwater marshes, tidal flats and dunes

the south-west Netherlands is one of ongoing struggle between people and the sea. People have been reclaiming the saltmarshes in this area and converting them into farmland since 1000 A.D. However, at various times the sea walls have been breached by storm floods and areas have reverted to the sea. On 1 February 1953, a northwesterly storm induced tides 3 m higher than normal; approximately 180 km of coastal defence dikes were breached and 160 000 ha of polder land was inundated. 1835 people lost their lives, more than 46 000 farms and buildings were destroyed or damaged and approximately 200 000 farm animals were lost (Fig. 12.2).

The Delta project, formalized in 1957 by an act of the Dutch parliament, was conceived as an answer to the permanent risk of flooding which threatens lives and property in this low-lying region. The core of the project was the closure of the main tidal estuaries and inlets in the south-west Netherlands. The Western Scheldt, however, was to remain open to allow international shipping to reach Antwerp; therefore the dikes along the river were raised (Huis in't Veld *et al.* 1984).

In order to be able to construct the primary sea walls in the mouths of the estuaries, the tidal current velocities in the estuaries had to be reduced. To do this, secondary compartmentalization dams were constructed (Fig. 12.3: 1,3 and 4), thereby reducing the area under intertidal influence. The Veerse Gat and Grevelingen estuaries (Fig. 12.3: 2 and 6) were closed off from the North Sea by high sea walls in 1961 and 1971 respectively, and turned into lakes. The Haringvliet (Fig. 12.3: 5) was closed in 1970 by large sluices intended to function as an outlet for the Rhine and Meuse.

The original plan for the Oosterschelde estuary was to build a 9-km-long dam across the mouth of the estuary, to be completed in 1978. This would have resulted in the tidal basin changing into a stagnant lake of Rhine water. However, because of a shift in public opinion, the final barrier differs drastically from the simple dam originally envisaged. Throughout the 1960s and early 1970s, nature conservationists and fishermen

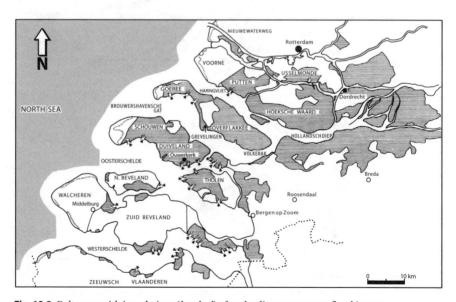

Fig. 12.2. Delta area with inundations (*hatched*) after the disastrous storm flood in 1953

conducted a campaign to raise awareness of the need to protect the area's outstanding natural resources and its unique tidal habitats, including the extensive shellfish industry (Smies and Huiskes 1981; Knoester 1984). This led to the Dutch government deciding to change the design of the dam in 1974. In 1976, after 2 years of desk studies, the Dutch parliament accepted a compromise solution: the storm-surge barrier (Fig. 12.3: 7). The barrier allows the tides to enter the estuary freely, thus safeguarding the tidal ecosystem, including the plant and animal communities and the shellfish cultivation. It also guarantees the safety of the lives and property of the human population when storm floods threaten the area. To maintain a tidal range of at least 2.7 m at a reference station (Yerseke) and achieve a tide-free shipping route from Rotterdam to Antwerp, compartment dams have been built at the northern and eastern boundaries (Fig. 12.3: 8 and 9) (Knoester *et al.* 1984; Nienhuis and Smaal 1994a).

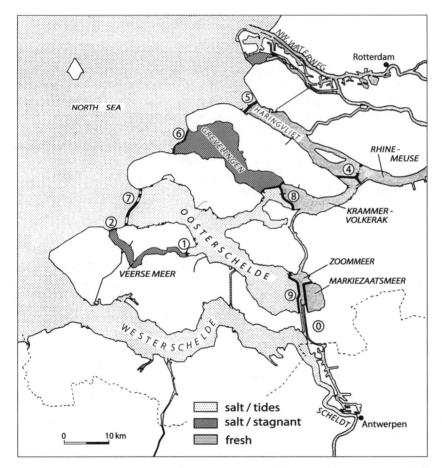

Fig. 12.3. Delta area with various water bodies resulting from the Delta project engineering scheme. *0* Kreekrakdam, 1867; *1* Zandkreekdam, 1960; *2* Veersegatdam, 1961; *3* Grevelingendam, 1964; *4* Volkerakdam, 1969; *5* Haringvlietdam, 1970; *6* Brouwersdam, 1971; *7* Oosterschelde storm-surge barrier, 1986; *8* Philipsdam, 1987; *9* Oesterdam, 1986. (After Nienhuis and Smaal 1994a)

The Delta project, particularly the storm-surge barrier, represents state-of-the-art storm-surge protection. The revolutionary designs and techniques, particularly with regard to foundation preparations and the use of foundation mattresses, reflected both innovative thinking and effective communication between scientists specializing in hydrodynamics and geomorphology and civil engineers (van Westen and Colijn 1984; Watson and Finkle 1990). The use of prefabricated components and the methods developed to install heavy constructions in open water influenced by waves and currents represent an important breakthrough in marine civil engineering. Moreover, this design represented a watershed in Dutch political decision making about the natural environment (Nienhuis and Smaal 1994a).

12.2
Responses of Ecology and Economy

12.2.1
Ecology

The Delta project has fundamentally changed the hydraulic, morphological and ecological characteristics of the region (Table 12.1). The overall change can be summarized as the loss of dynamic gradients (de Vries *et al.* 1996). The open and interconnected estuaries interfacing rivers and the sea have been replaced by isolated lakes and lagoons. Gradients have been replaced by discrete boundaries. New ecosystems with different characteristics developed within these new physical boundaries (Table 12.2). The factors controlling the newly created water systems are now biological and chemical (nutrients) rather than physical. A comparison of the Delta region before and after the Delta project reveals the following main change. (see Sects. 12.2.1.1–12.2.1.4)

12.2.1.1
Biodiversity

The total loss of 216 km^2 (59%) of tidal flats and 63.5 km^2 (68%) of salt marsh has significantly decreased characteristic estuarine habitats (Table 12.2) and has truncated salinity gradients. Part of the intertidal area has become a terrestrial habitat; another part is now under shallow freshwater. Increased transparency and the lack of fast cur-

Table 12.1. Present-day characteristics of former estuaries Haringvliet, Grevelingen, Oosterschelde which originally included Krammer-Volkerak, Veere and Westerschelde

	Haringvliet	Lake Grevelingen	Ooster-schelde	Lake Volkerak	Lake Veere	Wester-schelde
Tides	Stagnant	Stagnant	Tidal	Stagnant	Stagnant	Tidal
Salinity	Fresh	Saline	Saline	Fresh	Brackish	Saline
Trophic status	Eutrophic	Oligotrophic	Mesotrophic	Eutrophic	Eutrophic	Eutrophic
Polluted	Yes	No	No	Minor	Minor	Yes

rents in the stagnant lakes have promoted the development of benthic macrophytes. Herbivorous swans, geese and dabbling ducks have been attracted to these areas. However, the loss of intertidal areas has resulted in a significant decrease in shorebird populations (Meire *et al.* 1989; Schekkerman *et al.* 1994).

12.2.1.2
Productivity

Primary production in the Oosterschelde has been maintained because the phytoplankton community has adapted (Scholten and van der Tol 1994; Wetsteijn and Kromkamp 1994). As the growth of shellfish is significantly correlated with primary production, the carrying capacity for shellfish culture has also been maintained (Smaal and Nienhuis 1992; van Stralen and Dijkema 1994). Primary production is generally low in the Westerschelde estuary, owing to the high turbidity. In the stagnant eutrophic areas (Table 12.1) there are high densities of *Ulva* (Lake Veere) and microcystis (freshwater areas; de Hoog and Steenkamp 1989). The high nutrient turnover in the oligotrophic Lake Grevelingen supports a considerable primary production. Thus, the ecological

Table 12.2. Main abiotic characteristics of the Delta water systems and habitat changes resulting from the Delta project

	Haring-vliet	Lake Grevelingen	Ooster-schelde	Lake Volkerak	Lake Veere	Wester-schelde[c]	Total
Total area (km²)							
– Before	170	140	452[b]	65	40	300	1102[b]
– After	146[a]	108[a]	351	47	21[a]	300	926[a]
Tidal flats area (km²)							
– Before	38	63	183[b]	24	22	43	369
– After	2	0	118	0	0	33	153
Salt marsh area (km²)							
– Before	31	4	17[b]	6.5	7.5	35	94.5
– After	0	0	6	0	0	25	31
Flushing time (days)							
– Before	ND	ND	30	60	ND	60	
– After	1–>60	270	60	200	180	60	
Freshwater load (m³ s⁻¹)							
– Before	880	ND	70	50	ND	120	
– After	0–1300	5	25	30	3	120	
Average tidal amplitude (m)							
– Before	1	3	3.7	4	3	4	
– After	0.2	0	3.3	0	0	4	
Average depth (m)	5	5.4	7.8	4	4.2	ND	

ND No data
[a] Excluding new terrestrial areas
[b] Including the Volkerak area
[c] Impact not due to Delta project per se.

response to the Delta project in terms of biological productivity can be characterized as maintenance of functional stability (homeostasis), owing to adaptive and structural responses in the ecosystems.

12.2.1.3
Transformation Capacity

A descriptive model analysis comparing nitrogen loads and internal cycling for the four saline/brackish ecosystems (Lakes Veere and Grevelingen and Oosterschelde and Westerschelde) has revealed that the systems differ clearly in their transformation capacity, i.e. the potential for denitrification (de Vries *et al.* 1995, 1996). Nitrogen input in Lake Grevelingen has been balanced by removal of nitrogen by denitrification and retention in refractory detritus. The Oosterschelde shows more exchange with the North Sea, and denitrification balances the input by 60%. In these cases, denitrification is a function of organic matter cycling, which could be expressed as a cycling index (the ratio of uptake by primary producers and external input) of 7.1 for Lake Grevelingen and 3.7 for the Oosterschelde. The eutrophic Lake Veere has compensated input for 45% and the Westerschelde for only 25%; their cycling indices are 2.1 and 0.03 respectively. In these eutrophic systems denitrification is promoted by anaerobic conditions rather than organic matter cycling. The researchers concluded that the transformation capacity is underexploited in the water systems in the Delta area (de Vries *et al.* 1996).

12.2.1.4
Resilience

Resilience as a response parameter, defined as the ability to return to equilibrium after disturbance, can best be evaluated for the stagnant Lake Grevelingen. After the link with the sea was cut off in 1971, 20% of the total area was colonized by eelgrass (*Zostera marina*) which then unexpectedly declined sharply after 1989 and has almost vanished. The most probable cause of this dramatic decline is the increased salinity resulting from changes in the water management of this 'manipulated' lake, in combination with warm and dry summers and a reduction in silicate loads brought about by the diversion of water discharges (de Vries *et al.* 1996). Unexpected fluctuations have also been observed in other populations, particularly *Mytilus edulis*, *Ostrea edulis* and *Nassarius reticulatis*, (Lambeck 1982, 1985). This is evidence of a loss of resilience or robustness, i.e. increased vulnerability to external perturbation.

12.2.2
Socio-economy

The Delta project has changed the socio-economic functions of the water systems. As explained above, it was originally envisaged that all the estuaries except the Western Scheldt would be closed and freshwater lakes would be created. The overriding management aim was protection against storm floods; the freshwater lakes were intended to benefit agriculture. When the campaign to protect the area's outstanding natural resources and shellfish industry began in the 1970s, the sea walls closing the Grevelingen and Veerse Gat estuaries had already been completed, but work on the Oosterschelde

Table 12.3. Dominant socio-economic functions of the Delta water systems after the Delta project

Completion date	Haring-vliet 1970	Lake Gre-velingen 1971	Ooster-Schelde 1987	Lake Volkerak 1987	Lake Veere 1961	Wester-schelde n.a.
Agriculture	+++			++	++	
Recreation	++	+++	+		+++	
Shellfish culture		+	++			+
Nature conservation	+	++	+++	+		++
Shipping route				+++	+	+++

n.a., not applicable.

sea wall had just started. The government's decision to change the Delta project by building a storm-surge barrier in the Oosterschelde to maintain tidal influence (Nienhuis and Smaal 1994; van Westen and Colijn 1994) and to maintain saline conditions in the new Lakes Veere and Grevelingen had repercussions on the area's socio-economic functions. Instead of agriculture being the dominant function (owing to the availability of freshwater), the emphasis switched to recreation and nature conservation. Extensive recreational use had already developed in Lake Grevelingen because of the outstanding water quality. Nature conservation and shellfish cultivation became the main functions for the Oosterschelde estuary. Spatial differences in management aims now reflect temporal changes in the dominant socio-economic functions (Table 12.3). Recreational development has also been boosted by a former island becoming easily accessible because of the improved transport infrastructure.

Management has had to be intensified to maintain the various ecological and economic functions in the Delta waters. The man-made ecosystems have sometimes responded unexpectedly to management measures (Nienhuis and Smaal 1994b), demonstrating that they require special treatment. The shift in economic functions from agriculture to recreation, fisheries and nature conservation has raised new questions about the carrying capacity of such ecosystems for human use. New management strategies are required to resolve potential conflicts between these functions in a sustainable way.

12.3
Management Strategies

So far there have been three environmental management strategies in the Delta region:

- Reactive (single-issue) management, focusing solely on flood protection. This strategy, applied in the first phase of the project, aimed at total closure of the estuaries, thus transforming them into freshwater lakes. Feedback and buffering between coastal and inland waters were lost. This strategy was not oriented at sustainable development and increased the area's vulnerability to future catastrophes.
- Integrated management, focusing on various economic and ecological functions of the Delta region. This resulted in the preservation of existing valuable features in the

landscape and environment, the maintenance of saline conditions and the preservation of marshes by shore protection measures. The strategy was developed in response to the public demand for conservation and protection of the valuable features of tidal systems, particularly the Oosterschelde estuary. It led to a change in the water management of the Veere and Grevelingen lakes. The drawback of this focus on preserving existing features 'where they are now' is the need for ongoing intensive care because the natural ability to adapt has not been restored.

- Adaptive management based on functional properties of ecosystems and their interactions, as integrated elements of the landscape structure, and characterized by self-regulation and self-organization, i.e. homeostasis and resilient responses to perturbations. This strategy aims at environmental protection and at restoring and developing valuable features 'where they must be'. The development is intended to be sustainable, so estuarine gradients are being re-established by reintroducing tidal influence and by restoring tidal habitats. It is currently being decided how to restore tidal influence in the Haringvliet area, to increase the salinity gradient in the Oosterschelde and to restore intertidal habitats in the Westerschelde.

12.4
The Decision Making Process and Lessons Learned

As shown above, the aims of the Delta project have evolved from protection against storm floods to a whole suite of socio-economic aims. The actors in this process have been local people and the general public, politicians and government water managers (Fig. 12.4). The introduction of novel technology has been important in resolving conflicts between various socio-economic and ecological interests. In fact, this technology

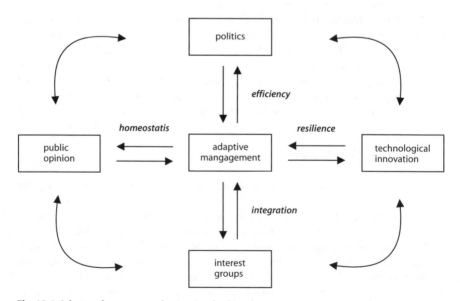

Fig. 12.4. Scheme of processes and actors involved in adaptive management strategies

enabled a series of breakthroughs in these conflicts. Future decisions on infrastructure will probably also rely heavily on new technology.

In the first stage (reactive management), coastal defence technology was developed to build sea walls in tidal areas, by using caissons and air transport to close off inlets. The innovation consisted mainly of the introduction of new techniques; there was no conflict of any importance. The public call for an open Oosterschelde resulted in a compromise between safety and other objectives (integrated management), with priority being given in the following order 1) environmental protection, 2) protection of the shellfish industry and 3) recreational functions. The compromise was a storm-surge barrier; the new technology required was a major challenge for coastal engineers and constructors. Management also had to be revamped, as other disciplines, socio-economic functions and public opinion became involved.

The public debate on the Delta region currently focuses on the more sustainable management of estuarine and enclosed systems. Although there is still much controversy, support is growing for the idea of habitat restoration and reintroduction of functional properties of estuarine systems, resulting in more resilient systems. Public involvement has increased and is now part of the decision making process. This type of approach requires adaptive management. The technology required for this type of management has not yet been determined. However, technological innovations can be expected to be inspired by the natural characteristics of the ecosystems in which the infrastructural measures are required.

Adaptive management can be characterized by (Fig. 12.4):

- The integration of various socio-economic and ecological functions, often identified by interest groups
- A homeostatic approach: exploiting self-regulation and self-organization – not only of ecosystems (see 4), but also of the socio-economic system. This requires the involvement of locals and of the general public in the process of deciding about future infrastructure
- The interaction between politics and management is based on efficiency and devolved responsibility: the approach is business-like
- A resilient approach: technological innovation with respect to infrastructure should exploit natural processes and resources

12.5
Conclusions

The Delta project resulted in the isolation of water systems and loss of estuarine gradients but also brought about changes in the socio-economic functions of the water systems: recreational use and nature conservation became dominant over agriculture. Continuous intensive care is required to maintain the newly created water systems. The current solution is an adaptive strategy for estuarine management which is based on the dynamic properties of estuarine systems. The key issue should be maintaining the connectivity between the upper reaches of rivers, their estuaries and the sea.

Acknowledgments. The authors are grateful to L. Bijlsma and I. de Vries for stimulating discussions and to J. Burrough-Boenisch for improving the English text.

References

de Hoog JEW, Steenkamp BPC (1989) Eutrophication of the fresh waters of the Delta. In: Hooghart JC, Posthumus CWS (eds) Hydro-ecological relations in the Delta waters of the south-west Netherlands: technical meeting 46, Rotterdam, The Netherlands. TNO committee on hydrological research, The Hague, pp 27–48

de Vries I, Phillipart CJM, de Groodt EG, van der Tol MWM (1995) Coastal eutrophication and marine benthic vegetation. In: Schramm W, Nienhuis PH (eds) Marine benthic vegetation. Recent changes and the effects of eutrophication. Springer Verlag Ecol studies 123:79–14

de Vries I, Smaal AC, Nienhuis PH, Joordens JCA (1996) Estuarine management strategies and the predictability of ecosystem changes. J Coastal Conservation 2:139–148

Huis in't Veld JC, Stuip J, Walter AW, van Westen JM (1984). The closure of tidal basins. Delft University Press, p 450

Knoester M (1984) Introduction to the Delta case studies. Wat Sci Tech 16:1–9

Knoester M, Visser J, Bannink BA, Colijn CJ, Broeders WPA (1984) The eastern Scheldt project. Wat Sci Tech 16:51–77

Lambeck RHD (1982) Colonization and distribution of *Nassarius reticulatus* (Mollusca: Prosobranchia) in the newly created saline lake Grevelingen (SW Netherlands). Neth J Sea Res 16:67–79

Lambeck RHD (1985) Leven zonder getij. In: Nienhuis PH (red) Het Grevelingenmeer, van estuarium naar zoutwatermeer. Natuur en Techniek pp 114–129 (in Dutch)

Meire PM, Seys J, Ysebaert T, Meininger PL, Baptist HJM (1989) A changing Delta: effects of large coastal engineering works on feeding ecological relationships as illustrated by waterbirds. In: Hooghart JC, Posthumus CWS (eds), Hydro-ecological relations in the Delta waters of the south-west Netherlands: technical meeting 46, Rotterdam, The Netherlands. TNO committee on hydrological research, The Hague, pp 109–145

Nienhuis PH, Smaal AC (1994a) The Oosterschelde estuary (The Netherlands): a case-study of a changing ecosystem. Kluwer Acad. Publ. p 597

Nienhuis PH, Smaal AC (1994b) The Oosterschelde (The Netherlands), an estuarine ecosystem under stress: discrimination between the effects of human-induced and natural stress. In: Dyer KR, Orth RJ (eds) Changes in fluxes in estuaries. Olsen and Olsen, Denmark, pp 109–120

Saeijs HLF (1982) Changing estuaries, a review and new strategy for management and design in coastal engineering. RWS communications 32:414

Schekkerman H, Meininger PL, Meire PM (1994) Changes in the waterbird populations of the Oosterschelde (SW Netherlands) as a result of a coastal engineering project. Hydrobiologia 282/283:509–524

Scholten H, Klepper O, Nienhuis PH, Knoester M (1990) Oosterschelde estuary (SW Netherlands): a self sustaining ecosystem? Hydrobiologia 195:201–215

Scholten H, Van der Tol MWM (1994) SMOES: a simulation model for the Oosterschelde ecosystem. Part II: calibration and validation. Hydrobiologia 282/283:453–474

Smaal AC, Nienhuis PN (1992) The eastern Scheldt (The Netherlands), from an estuary to a tidal bay: a review of responses at the ecosystem level. Neth J Sea Res 30:161–173

Smies M, Huiskes A (1981) Holland's eastern Scheldt estuary barrier scheme: some ecological considerations. Ambio 10:158–165

van Stralen MR, Dijkema RD (1994) Mussel culture in a changing environment: the effects of a coastal engineering project on mussel culture in the Oosterschelde estuary (SW Netherlands). Hydrobiologia 282/283:359–380

van Westen CJ, Colijn CJ (1994) Policy planning in the Oosterschelde estuary (SW Netherlands). Hydrobiologia 283/283:563–574

Watson I, Finkle CW (1990) State of the art in storm-surge protection: The Netherlands Delta project. J Coastal Res 6:739–764

Wetsteyn LPJM, Kromkamp JC (1994) Turbidity, nutrients and phytoplankton primary production in the Oosterschelde estuary (SW Netherlands) before, during and after the construction of a storm-surge barrier. Hydrobiologia 282/283:61–78

CHAPTER 13

Economics and Project Management in the Coastal Zone

R. Kerry Turner

13.1
Introduction

The proliferation of large-scale construction projects in coastal zones reflects the increasing environmental pressures being exerted on such areas (both biophysical and socio-economic systems) by population growth/migration, urbanisation, industrial development and the globalisation of economic/trading activities, as well as tourism. Around 37% (2.07 billion) of the 1994 world population (of 5.62 billion) live within 100 km of a coastline and 44% within 150 km of a coastline (Cohen *et al.* 1997). These trends seem set to continue.

The aggregate effect of these pressures has been to generate multiple and often conflicting resource usage trends in coastal zones. The mitigation of these resource conflict problems through the practical adoption of the sustainable development policy objective, which politicians have signed up to, requires both underpinning comprehensive resource assessment (including resource valuation) methods/techniques and institutional procedures such as integrated coastal zone planning and management (Bower and Turner 1998).

Since the 1960s the use of cost benefit-analysis (CBA) and its attendant monetary valuation methods and techniques has been extended to cover a wide spectrum of projects, policies and programmes. This has inevitably led to relative valuations of a diverse set of goods and services, including environmental costs and benefits (Table 13.1).

Table 13.1. Spectrum of appraisal methods. (Pearce and Turner 1992)

Financial appraisal	Economic appraisal	Multi-criteria approach
Based on private costs and benefits in cash flow terms	Based on social costs and benefits, expressed in monetary terms, including environmental effects	Based on non-monetary and monetary estimates of a diverse range of effects, social, political and environmental
Analysis is related to an individual economic agent, i.e. farmer, householder, firm or agency	Social costs/benefits = private costs/benefits + external costs and benefits	Scaling and weighting of impacts
Typical techniques: discounted cash flows and balance sheets; payback periods and internal rates of return	Typical techniques: cost-benefit analysis, extended cost-benefit analysis, and risk-benefit analysis	Typical techniques: impact matrices, planning balance sheets, concordance analysis, networks, and trade-off analysis
← less comprehensive/less data intensive		more comprehensive/more data intensive →

Table 13.2. Environmental evaluation methods, showing increasing complexity and scale of analysis (Pearce and Turner 1992)

Financial analysis	Economic cost-benefit analysis	Extended cost-benefit analysis	Environmental Impact Assessment	Multi-criteria decision methods
Financial profitability criterion; Private costs and revenues; Monetary valuation	Economic efficiency criterion; Social costs and benefits; Monetary valuation	Sustainable development principles; Economic efficiency and equity trade-off; Environmental standards as constraints; Opportunity costs analysis	Quantification of a diverse set of effects on a common scale, but no evaluation; or misleading composite index scores	Multiple decision criteria; Monetary and nonmonetary evaluation combinations

Standard CBA has at its core the economic efficiency criterion, but the real world political economy mirrors the interests of the different stakeholders in society. Multiple decision criteria, efficiency, equity, environmental sustainability, etc., compete for priority status across the different decision contexts within the process that determines the net social welfare gains and losses involved. Since environmental capital resources (natural goods and services) are undoubtedly subject to scarcity the economic efficiency criterion cannot be ignored in rational decision making. However, its claim to represent a meta criterion is both undesirable in social welfare terms and politically naive.

The polar opposite positions that have evolved as CBA has infiltrated the policy process are both suspect. The standard economic efficiency position unrealistically forces all of socio-economic activity and life into the market calculus (individual willingness to pay or be compensated values). Some supporters of this world view also seem to believe that the analytical constructs used in the assessment of projects and programmes yield a value-free outcome. This is misleading and works against participatory planning and resource management. Value judgements are inevitable and should be made transparent not shrouded in technical analytical complexity. CBA has great power as an heuristic aid to the policy process, but only if its central economic welfare assumptions are clearly visible.

When they are not visible the danger of "regulatory/institutional" capture is increased.

13.2
Competing Analytical Frameworks and Underlying World Views

13.2.1
Conventional CBA Framework

In the conventional CBA framework approach, conventional CBA is retained as a narrow measure of the economic efficiency of alternative projects and/or courses of action. Equity considerations would not be addressed within the economic analysis. The scope of benefits estimation would be limited and a number of environmental effects would be excluded because of the lack of market price data or easily accessible direct proxy price data.

Multi-criteria analysis and/or Environmental Impact Assessment (EIA) would be seen as entirely separate from CBA. They might contribute to an alternative environmental appraisal process not formally linked with CBA. Environmental effects in this interdisciplinary process would be quantified but not translated into monetary terms. Decision-makers would be presented with two or more project/policy appraisal reports, only one of which would contain monetary net benefit calculations. These calculations would be restricted to those impacts directly amenable to monetary valuation and would be based on the single criterion of economic efficiency.

13.2.2
Modified CBA Framework

In the modified CBA framework approach, conventional CBA is extended to encompass the sustainable economic development policy objective. Projects and/or courses of action would be appraised on the basis of both economic efficiency criterion and an equity criterion designed to monitor the distribution of costs and benefits. Both the immediate and the long-term distribution would be assessed (i.e. intragenerational and intergeneration equity). Benefits estimation would be utilised as extensively as possible, incorporating as many environmental effects as was practicable given the prevailing state of the art of monetary valuation methods and techniques.

Distributional issues and environmental resource use inefficiencies (sustainability constraints) would be responded to not by discount rate manipulation or the abandonment of discounting, but by the use of compensating offsets (shadow projects). Where feasible and practicable, compensation for the loss of or damage to environmental assets would be actual and not merely hypothetical. Shadow projects would be identified and fully costed. These compensation costs would then be added to the development project's costs and included in the cost-benefit calculation. Magnitude and significance of environmental impacts would then be standardised in monetary terms, as far as was practicable, by economic valuation methods and techniques.

In cases where compensating offsets for environmental asset loss or damage were not available, the social opportunity costs of foregoing the development project's net benefits would need to be carefully considered against a range of criteria.

13.2.3
Radical Restriction of CBA Framework

The central concept in the radical restriction of CBA framework approach is the preemptive environmental standard imposed by regulatory means. These prior environmental constraints on economic development would take the form of an extensive array of environmental ambient quality standards, conservation zones and national parks, nature reserves and other designated protected sites, etc. The standards would be set prior to any economic analysis of the resource issues involved. EIA would play a lead role in this standard-setting phase in terms of bio-physical and some socio-economic data gathering and impact measurement and prediction. Political, cultural and ethical criteria would all be deployed in the standard setting process.

Economic analysis would be restricted to cost-effectiveness analysis, designed to identify the least costly ways of implementing and enforcing the environmental standards. Economic analysis would not be deployed to explicitly test the worthwhileness of the standards themselves. However, once the standards had been set (by whatever means) implicitly relative price/valuations would also have been determined.

13.2.4
Abandonment of CBA Framework

In the abandonment of CBA framework approach (the deep Ecology approach), the primary policy objective becomes maximum nature conservation. Nature is seen as possessing inherent value and species other than humans would be viewed as possessing interests and rights. Zero-economic growth strategies and minimum resource-take economic systems (small-scale, decentralised, based on organic agriculture and renewable energy sources) would be the order of the day. There would be no role for benefits estimation and economic analysis, in their conventional mode, in this approach.

13.2.5
Further Issues of CBA Framework

Equally flawed is the world view which calls for the complete abandonment of CBA in favour of a set of ill-defined alternative assessment frameworks, e.g. energy analysis, life cycle analysis, bio-ethical imperatives, etc. Economic realities are somehow pushed aside and spurious claims to the moral 'high ground' often substituted in the name of environmental imperatives.

The middle ground offers more prospects for both constructive analytical debates and better enabling methods and techniques leading to improved resource assessment and social policy making. Debate is required because a number of difficult issues and questions remain to be resolved. The acceptance of sustainable development as a policy objective will serve to heighten the tensions between the polar extreme camps, while simultaneously offering an important policy context for other analysts to pursue a joint and interdisciplinary research agenda (see Sect. 13.2.1–13.2.4).

Future progress contributing to the achievement of a socially acceptable 'sustainability balance' [trading off efficiency, equity and precautionary principles $(E:E:P)$] is required in the following areas:

- Defining the boundaries of the domains for the individualistic market-based calculus and economic welfare value judgements and non-market collectivist values and calculus
- Further refinement of monetary valuation methods such as contingent valuation
- Development/refinement of non-market assessment methods and techniques
- Formulation of an integrated set of multi-criteria heuristic analytical aids which can provide for more participatory decision-making

This chapter focuses on the resource assessment issues and in particular on the evaluation of the social costs and benefits surrounding large-scale capital investment projects.

13.3
Large-Scale Project Evaluation in the Coastal Zone

A diverse range of existing and planned projects covering, among others, energy extraction and supply, transport facilities and infrastructure, sea defence and coastal protection schemes, and tourism-related housing and recreation facilities are or will be located at or near the coast. It does not seem possible to formulate a priori a comprehensive typology of such projects in terms of the magnitude and significance of their environmental and socio-economic welfare impacts. The aggregate impact of such projects is both context – and spatially specific. On the other hand, it is possible to construct a general economic – ecological assessment methodology which is capable of providing guidance to coastal resource managers and policymakers.

Because policy decisions are required relating to a range of spatial and temporal scales and different socio-economic and political levels, several broad assessment categories (ignoring the time dimension for the moment) need to be distinguished (Barbier 1993). It might be the case that a given existing project imposes a particular impact on an individual coastal resource or set of resources, e.g. heavy metal discharge from an industrial plant, oil spillage from platforms, storage facilities or during transport, sewage disposal from urban areas. Thus, in this *impact analysis* category a specific environmental impact is assessed via the valuation of the environmental state changes in the coastal resource(s) connected to the impact. The valuation requires an estimate of the consequent net coastal resources production and environmental benefits effects. The total cost of the impact (P_c) in social welfare terms is the forgone net benefits (NB_{fe}); so $P_c = NB_{fe}$. The forgone net environmental benefits related to a pollution impact, for example, can then be compared with a range of alternative pollution abatement options and their cost (product/process design modifications for

Table 13.3. Coastal environmental impacts and valuation methods. (Adapted from Turner and Adger 1996)

Effects categories	Valuation method options
Productivity E.g. fisheries, agriculture, tourism, water resources, industrial production, marine transport, storm buffering and coastal protection	Market valuation via prices or surrogates Preventive expenditure Replacement cost/shadow projects/cost-effectiveness analysis Defensive expenditure Human capital or cost of illness
Health	Contingent valuation Preventive expenditure Defensive expenditure Contingent valuation/ranking
Amenity	Travel cost method
Coastal ecosystems, wetlands, dunes, beaches, etc., and some landscapes, including cultural assets and structures	Hedonic property method
Existence values Ecosystems; cultural assets	Contingent valuation

waste minimisation, end-of-pipe treatment and 'safe' disposal, etc). Table 13.3 summarises some relevant environmental state changes and related economic valuation methods.

A second assessment category, partial valuation, encompasses situations which require the evaluation of alternative resource allocations or project options. A planned large-scale project (or extension of an existing project) such as a residential/recreational housing complex or port and harbour facilities, might require the conversion of coastal wetlands and mudflats with significant biodiversity and other functional values. So the net benefits of the wetland conversion (NB_c) would be the direct benefits of the project (B_D), minus the direct costs of the project (C_D = capital and operating costs), minus the forgone net production and environmental benefits of the conserved wetland (NB_{fe}):

$$NB_c = B_D - C_D - NB_{fe} > 0 .$$

It is sometimes the case that estimation of only some elements of the valuation expression above is necessary to prove that the development project is uneconomic, provided that the on–going utilisation of the natural system is at a sustainable level. An analysis of the opportunity cost of wetland conservation (i.e. forgone project direct net benefits), for example, might show that $B_D - C_D$ is only marginally positive. Some past agricultural conversion schemes in Europe and housing developments in the USA have actually been shown to yield negative opportunity costs (Turner *et al.* 1983; Batie and Mabbs-Zeno 1985).

As long as the conserved wetland yields a flow of functional benefits, storm buffering capacity, fish and other product outputs etc., the positive valuation of only some of these outputs/services will sometimes be enough to tip the economic balance against the large-scale project. On the other hand, the development project may generate significant employment and regional income benefits and be seen as part of a regional development policy strategy. Increasing employment/reducing regional income disparities may therefore be interpreted as pre-emptive constraints on the cost-benefit analysis and such benefits may be heavily weighted by policymakers.

A third assessment category covers the evaluation of protected areas schemes involving restricted or controlled resource use. Such marine park or coastal nature reserve schemes, for example, might be a required compensating shadow project element in a large-scale project programme approval process or alternatively might preclude the existence of any given project altogether. The precise circumstances will depend on how 'weakly' or 'strongly' sustainability standards/constraints are interpreted and imposed by planning/management authorities (Turner 1993). The on-going loss of coastal wetlands might have reached such a stage that regulatory authorities were seeking to impose a 'no net wetland loss' rule on all future development activity in the coastal zone (a pre-emptive environmental policy constraint on CBA).

In situations where there is a direct choice between a development project and a marine park or similar conservation scheme, or where compensating environmental shadow project possibilities are not available, it may be necessary to use the total valuation approach. The analysis would seek to determine whether the total net benefits of the protected area kept in a sustainable 'natural' state (NB_p) exceeded the direct costs of establishing the protected zone and necessary buffer zone (C_p), plus the net benefits forgone (NB_{fd}) of alternative development uses of the protected area. The conservation zone plus buffer zone setup costs may include costs of relocating or compensating existing users:

$$NB_p - C_p - NB_{fd} > 0 \ .$$

Finally, a hypothetical project (port, harbour and recreational housing and marina complex) is used to illustrate the partial and total valuation methods on the basis of the economically efficient resource usage principle over time (with and without uncertainty) (see Sect. 13.3.1; Pearce and Turner 1992). The deployment of sustainability constraints/standards is also analysed in the same project choice context.

13.3.1
Port/Marina Complex Project: Partial Valuation Approach

Imagine that a port and marina project is scheduled and that it is to be sited so that it displaces an existing and environmentally valuable wetland area. Let us further assume that the project will also generate pollution impacts related to sewerage effluent disposal and other waste emissions/discharges. The basic cost-benefit decision rule (on the basis of the economic efficiently criterion) is:
Accept a proposal if over time

$$\int t \{ B_{Dt} - C_{Dt} - NB_{fet} - TEC_t \} e^{-rt} > 0 \ , \tag{13.1}$$

where B_{Dt} = direct project benefits, C_{Dt} = direct project costs, NB_{fet} = forgone net production and environmental benefits of the conserved wetland, TEC_t = pollution damage costs (net effects on coastal resources' production and environmental benefits) and r = discount rate.

In a deterministic world in which there is no uncertainty, we need to estimate the *total economic value* (*TEV*) of NB_{fe} and *TEC*:

$$TEV = TUV + TNUV \ ,$$

where TEV = total economic value, TUV = total use value (relating to wetland and other coastal resources impacted by pollution) and $TNUV$ = total non-use (existence) value (see Fig. 13.1). If this TEV is sacrificed because of the development project, then the basic cost-benefit decision rule [Eq. (13.1)] becomes: Accept if

$$\int t \{ B_{Dt} - C_{Dt} - [TUV + TNUV] \} e^{-rt} > 0 \ . \tag{13.2}$$

In circumstances where there is uncertainty, for example, supply uncertainty about the continued availability of the environmental asset in question, but consumers will almost certainly demand the environmental goods/services in the future, then the concept of option value (*OV*) is relevant. Option value is the difference between the amount someone is willing to pay in order to retain the option of future use of an environmental asset (the option price) and the expected value of the ex post consumer surplus

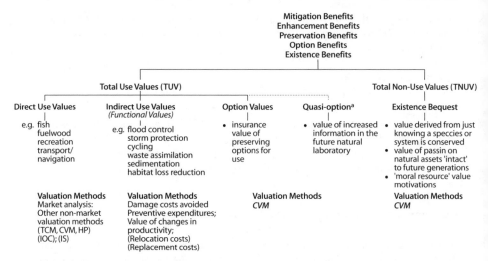

Fig. 13.1. Methods of valuing coastal zone benefits. (Adapted from Turner 1988 and Barbier 1989) Notes: *Market analysis* based on market prices; *HP* hedonic pricing, based on land/property value data; *CVM* contingent valuation method based on social surveys designed to elicit willingness to pay values; *TCM* travel cost method, based on recreationalist expenditure data; *IOC* indirect opportunity cost approach, based on options foregone; *IS* indirect substitute approach; *TEV* total economic value. Benefits categories illustrated do not include *indirect* or *secondary benefits* provided by the coastal zone to the regional economy, i.e. the regional income multiplier effects. [a] not commensurate with other elements of TEV.

(equivalent of wetland TUV). Equation (13.2) is adjusted for uncertainty by the addition of OV:

$$\int t\{B_{Dt} - C_{Dt} - [TUV_w + OV - TNUV_w] - TUV_{CR}\}e^{-rt} > 0 \ , \tag{13.3}$$

where TUV_w = total use value of the displaced wetland, $TNUV_w$ = total non-use value of the displaced wetland, TUV_{CR} = lost total use value related to the pollution of the coastal zone resources and $TNUV_{CR}$ = lost total non-use value due to pollution in the coastal zone.

13.3.2
Sustainable Development Policy Objective

In broad terms, sustainable development involves providing a bequest from the current generation to the next of an amount and quality of wealth which is at least equal to that inherited by the current generation. This requires a non-declining capital stock over time and is consistent with the intergenerational equity criterion. Sustainability therefore requires a development process that allows for an increase in the well-being of the current generation, with particular emphasis on the welfare of the poorest members of society, while simultaneously avoiding uncompensated and 'significant' costs to future generations. Policy would be based on a long-term perspective, would incorporate an equity as well as an efficiency criterion, and may also emphasise the need to maintain a 'healthy' global ecological system.

'Sustainable Coastal Development' can therefore be defined as the proper use and care of coastal environmental systems borrowed from future generations. Because of the existing and likely continuing heavy usage pressures on coastal areas, a principle of sustainable utilisation of resources is a reasonable guiding concept. The resource base includes produced capital, human capital and natural capital assets such as raw materials, waste receptors, landscape and amenity assets.

The 'constant capital' condition for sustainable development can be interpreted in a weak and strong form. The weak sustainability condition can be written as:

$$K / N = \{K_m + K_h + K_n + K_{sm}\}/ N .$$ (13.4)

Equation (13.4) should be constant or rising over time. The strong sustainability condition in its environmental form can be written as:

$$K_n / N .$$ (13.5)

Equation (13.5) should be constant or rising over time *and* weak sustainability [Eq. (13.4)] must also hold. K_m = man-made capital, K_h = human capital, K_n = natural capital, K_{sm} = social/moral capital and N = population.

Weak sustainability effectively assumes unlimited substitution possibilities (via technical progress) between the different forms of capital. Strong environmental sustainability assumes that natural capital (or 'critical' components of such environmental systems) cannot be substituted for by other forms of capital.

Because the coastal zone is the most biodiverse zone, a strong sustainability strategy would impose a 'zero net loss' principle or constraint on resource utilisation (affecting habitats, biodiversity and the operation of natural processes). Wetlands provide a range of valuable functions and related goods/services flows. Such systems have also been subject to severe environmental pressures and have suffered extensive degradation and destruction. They may therefore be good candidates for a 'zero net loss' rule depending on how critical the functions and systems involved might be. The opportunity costs of the wetland conservation policy (i.e. foregone development project net benefits) should be calculated and presented to policymakers. If the wetland area requires a more proactive management approach, i.e. buffer zone creation, monitoring and enforcement costs, then the total valuation calculation will be required.

13.4
Programme Level Sustainability Rules

Instead of just concentrating on single projects it is possible to take a more comprehensive and strategic approach across a set of projects throughout the coastal zone and connected drainage basin. The constant natural capital rule at this programme level can be interpreted as a process of netting out environmental damage costs ($NB_{fet} + TEC_t$) across a set of projects, such that the sum of individual damages should be zero or negative (Barbier *et al.* 1990):

$$\sum_i E_i \leq 0 ,$$

where E_i = environmental damage ($NB_{fe} + TEC$) generated by the ith project.

Under a strong sustainability rule, $\sum E_i$ is constrained to be non positive for each period of time. If it is not feasible for E_i to be zero or negative for all projects, it may be possible to include within any portfolio of projects one or more shadow projects. These shadow projects aim to compensate for the environmental damage generated by the existing/planned set of projects, and are not subject to normal cost-benefit rules. Thus, the loss of a wetland at some particular location may be compensated for by wetland relocation, creation or restoration investments elsewhere in the zone.

Environmentally compensating project(s), j, would be chosen such that for strong sustainability,

$$\sum_j A_{jt} \geq \sum E_{it}, \forall_t \ ,$$

where A_j = net environmental benefits of jth project.

13.5
Conclusions

Given the diversity of large-scale capital investment projects that have been, and continue to be, constructed in coastal zones there is an urgent requirement to assess both the social costs and benefits of such environmental pressure in naturally variable and often sensitive coastal ecosystems. The cost-benefit approach outlined in this chapter provides one such decision-aiding framework and method. It is flexible enough to cover the complete scale spectrum, from individual plants/projects up to marine conservation zoning. It is also possible to adapt the approach to include sustainability and/or other policy objective constraints. What it should not be interpreted to be is a panacea for difficult E : E : P trade-off decisions and a technical substitute for the democratic political process.

References

Batie SS, Mabbs-Zero CC (1985) Opportunity costs of preserving coastal wetlands: a case study of a recreational housing development, Land Economics, 61:1–9
Bower BT, Turner RK (1998) Characterising and analysing benefits from integrated coastal management, Ocean and Coastal Management (in press)
Barbier EB (1989) The economic value of ecosystems: tropical wetlands. LEEC Gatekeeper Series 89–02, London Environmental Economics Centre, London
Barbier EB (1993) Sustainable use of wetlands: valuing tropical wetland benefits. The Geographical Journal 159:22–32
Barbier EB, Markandya A (1990) The conditions for achieving environmentally sustainable development. European Economic Review 34:659–69
Cohen JE et al. (1997) Coastal populations. Nature 278:1211–12
Pearce DW, Turner RK (1992) Benefits estimates and environmental decision-making. OECD, Paris
Turner RK (1988) Wetland conservation: economics and ethics. In: Collard D et al. (eds) Economics, growth and sustainable environments. Macmillan, London, pp 121–159
Turner RK (1993) Sustainable environmental economics and management: overview. In: Turner RK (ed.) Sustainable environmental economics and management: principles and practice. Belhaven, London, chapt. 1
Turner RK, Adger NA (1996) Land-ocean interactions in the coastal zone (LOICZ), coastal zone resources assessment guidelines. LOICZ/IGBP Reports and Studies No. 4, Texel, Netherlands and Stockholm
Turner RK, Dent D, Hey RD (1983) Evaluation of the environmental impact of wetland flood protection and drainage schemes. Environment and Planning A, 15:1777–96

CHAPTER 14

An Integrated Approach to Sustainable Coastal Management

Christina von Schweinichen

14.1
Introduction

There is no common definition of what constitutes a coastal zone, but rather a number of complementary definitions, each serving a different purpose. Although it is generally intuitively understood what is meant by "the coastal zone", it is difficult to place precise boundaries around it, either landward or seaward. Landward boundary of the coastal zone is particularly vague; some oceans can affect climate far inland from the sea. The coastal zone is the zone in which most of the infrastructural and human activities directly connected with the sea are located (European Environment Agency 1995, p 568).

Human activities are often concentrated in coastal regions and not always suited to assimilating those activities, and where adverse effects are most apparent. Coastal zones are relatively fragile ecosystems, and disordered urbanization and development of infrastructure alone or in combination with uncoordinated industrial, tourism-related, fishing, and agricultural activities, can lead to rapid degradation of coastal habitats and resources. Mounting pressure on the coastal zone environment has, in several Economic Commission for Europe (ECE) member countries, resulted in a rapid decline in open spaces and natural sites and in a lack of space to accommodate coastal activities without significant harmful effects. ECE work is carried out by searching for new directions in reconciling the promotion of economic growth on the one hand, and safeguarding the heritage of ECE member countries from a physical, social and environmental point of view on the other.

In this context it is worth mentioning that the ECE has promoted the *Guidelines on Sustainable Human Settlements Planning and Management* (UN ECE 1996a) with the aim of providing alternatives to translate the overall sustainable development policies into operational planning criteria. In particular it has looked at opportunities to broaden national experiences and practices of translating ideas regarding sustainable planning and management into practice.

14.2
Management of Coastal Areas

Coastal management requires a special priority in ECE countries as it experiences significant pressures from development. Europe only has a coastline of 143 000 km, with a great number of islands known for their particular and sensitive environment (ibidem). Specific environmental threats to coastal zones relate to increase of urban activities and subsequently damage in land-use patterns, sea level rise, changes in hydrological cycles of major rivers, pollution including contamination through waste dumping, loss of habitats and erosion, etc.

The main aim should be to prevent damage to coastal areas by identifying specific goals for sustainability, developing and implementing strategic planning and designing instruments to achieve these goals. Planning of coastal areas has to deal with the task of managing the spatial consequences of a wide range of technological, social and environmental developments, to guarantee both continued economic and social progress and environmental sustainability.

An integrated approach to monitoring and managing coastal zones could help to overcome the difficulties which arise from the different uses of coastal zones; addressing the causes of environmental development problems, instead of focusing on their symptoms, is a necessary condition for success. Moreover, the interrelated nature of coastal environmental problems requires that actions undertaken at various levels should be part of an integrated approach. It would bring together national and local bodies with legislative and regulatory responsibilities for specific tasks in marine and coastal zone management.

An ECE investigation of how and to what extent the *Guidelines on Sustainable Human Settlements Planning and Management* are reflected in the national planning was conducted with a group of countries (Implementation of sustainable human settlements policies: the case of Finland, Greece, Norway and Poland). Recent developments indicated that in the institutional hierarchies, the role of local authorities is emphasised. The delegation of planning power to the local level varies in national contexts, as does the scope for public participation and voluntary action facilitated by the regulations. Problems in the functioning of the planning system, however, are mostly associated with local conflicts over land-use allocation, efficient control over development and trade-offs between conflicting objectives associated with economic, social and environmental priorities. Making environmental concerns a top priority is, however, problematic. The experience indicates that strong industrial interests do not want to be "fenced in" by local planning. Likewise, the municipalities tend to interpret national goals according to their own priorities. Various local issues compete with national principles and often local pragmatism tends to overrule the environmental expertise.

The ECE *Guidelines on Sustainable Human Settlements Planning and Management* (UN ECE 1996a) and the *Environmental Programme for Europe* (UN ECE 1995a) recommend promoting at the national level of close coordination on environmental aspects when preparing major sectoral decisions. Planning policies affecting the individual elements of each sector, such as housing, transport and industry, and infrastructure have direct consequences in terms of land consumption, loss of natural areas (habitats and species), energy consumption, effectiveness of infrastructure investment, quality of air, soil and water, stability of natural processes (e.g. water filtration, climate, cleansing air of pollutants, etc.), equitable access, range of options or choices, and human health, safety, mobility and enjoyment.

The environmental problems addressed in the ECE *Guidelines* are reflected in national legislation to a different extent. Although none of the countries closely looked at (*ibidem*) lacks environmental legislation, the actual priorities and national strategies vary substantially. The reason for this diversity is in part related to differences in political and economic priorities. Environmental issues are rarely a top priority. Rather, an "all-win" situation seems to be pursued. Often a difficult and delicate balance between costs and benefits and their societal distribution is achieved. In none of the countries do environmental concerns seem to have an absolute say in development issues. However, differences in compliance with environmental goals are also attributed to the notable variations in the organizational and administrative framework of policy implementation. The implementation of environmental policies and regulations involves different ministries and public agencies. Although national policies increasingly seem to address global and local environmental concerns, these policies are still too general, reflecting political ambitions and rhetoric rather than commitment and action.

The need to integrate environmental and developmental decision-making processes is a basic and recurrent principle stressed in several documents. In particular, *The Rio Follow-Up at Regional Level* (UN ECE 1993) recommends that the sustainable development process should be systematically monitored and evaluated, and the state of the environment regularly reviewed, with the aim of ensuring the transparency of and accountability for the environmental implications of economic and sectoral policies, including for the coastal areas.

The implementation of the methodology used in the 1991 Espoo Convention on Environmental Impact Assessment (EIA) in a Transboundary Context lays an important foundation in that respect. EIA has become a major tool for an integrated approach to the protection of the environment, since it requires a comprehensive assessment of the impact of an activity on the environment. Moreover, it looks into alternatives to the proposed activity and brings facts and information on environmental impacts to the attention of the decision makers and the public. It is important to ensure that EIA is applied to policies, plans and programmes at an early stage. The Convention is the first multilateral treaty to specify the procedural rights and duties of Parties with regard to the transboundary impacts of proposed activities. *Appendix I* to the Convention covers 17 groups of activities to which the Convention applies, such as nuclear and thermal power stations, road and railway construction, chemical installations, waste-disposal facilities, oil refineries, oil and gas pipelines, mining, steel production, pulp and paper manufacturing, construction of dams and reservoirs, trading ports, and offshore hydrocarbon production. All these activities are of high relevance for coastal areas. EIA is already used as an effective instrument at the national level and it is understood that the ECE Convention on Environmental Impact Assessment in a Transboundary Context (E/ECE/1250) will lead to environmentally sound and sustainable development by providing information on the interrelationship between economic activities and their environmental consequences in a particular transboundary context.

During the last 15 years, four conventions and five protocols have been developed in the ECE on air pollution, environmental impact assessment, industrial accidents and transboundary waters. The importance of these legal instruments as effective tools to promote active, direct and action-oriented international cooperation at the regional and sub-regional level is growing in view of the Commission's increasing membership, the many new borderlines cutting through Europe and, hence, the growing potential for transboundary environmental problems. These treaties, which are important elements of a common European legal framework, are concrete and effective instruments to eliminate the former dividing line between east and west and to integrate countries with economies in transition into a pan-European legal and economic space.

14.3
The Ecosystem Approach to Land-Use Planning in Coastal Areas

The goal of sustainability has affected current planning approaches and practices. The need to restore and maintain ecological health, reconsider patterns of development to prevent ecological degradation and enhance sustainable economy has led to the notion of ecosystem-based planning. Thinking about ecosystems in the context of planning and management represents a major shift from traditional land-use planning. An ecosystem is an interacting system of air, land, water and living organisms: any individual organism is an integral part of the system.

The key to understanding the ecosystem is to recognize that everything is connected to everything else and that, therefore, human activities should be viewed as interacting with pre-existing conditions and contributing to cumulative changes in ecosystem health. The ecosystem approach recognizes the dependence of human communities and economic systems on healthy environment including clean air, land and water, renewable and non-renewable resources, natural areas and wildlife.

As an example, the "dual network strategy" (van der Wal 1993) was developed in The Netherlands. This is an ecologically inspired planning method in which the transport network is seen as having a guiding effect on highly dynamic uses, such as business, offices, mass recreation and agriculture, while the water network influences less dynamic uses such as water collection, nature and low key recreation. The combination of both networks offers a framework to guide settlement developments and enables locations to be identified.

The ecosystem approach can be used on any scale: regional, municipal or urban. In its essence it emphasizes ecosystem health, sustainability and quality of life. Furthermore, it focuses on interactions between components of the different ecosystems, taking a long-term perspective. In determining coastal area development and design, key elements of environmental assessment process are incorporated, including examination of alternatives and prediction of effects. Planning is undertaken by all stakeholders in the development process.

The global implications of sustainability are immense. Planning for sustainability means becoming more aware of the connection between land use, resource use, waste generation, pollution of air, soil and water as well as social and economic impacts. Within the context as a whole, ecosystem growth and development must be assessed to ensure the protection of environmental and community values. Proposals and plans should be judged not only on economic merits, but also on their contribution to human activities as well as social objectives. Increasingly, the notion of expected contribution to the regeneration and sustenance of ecological health of the region should become a primary concern.

14.4
Capacity Building, Major Groups and Public Participation

Success in improving urban environmental management comes from the participation of all parties and the public in the process. High levels of government should set long-term priorities, finance major infrastructure developments and other initiatives to alleviate the environmental impact of human activities, design and legislate on environmental taxes, fees and standards, etc.

On the other hand, local authorities construct, operate and maintain economic, social and environmental infrastructure, oversee planning processes, establish local environmental policies and regulations, and participate in implementing national and regional environmental policies. They play a vital role in educating, mobilizing and responding to the public to promote sustainable development also in the coastal zones.

Furthermore, other parties are involved in the process besides central and local government: planners and administrators, the business sector and corporations, non-governmental and community organizations, the academic community, and local interest groups. Representatives of associations of cities and other local authorities should also increase levels of cooperation. One of the fundamental prerequisites is broad public participation in decision-making. This includes the need for individuals, groups and organizations to participate in EIA procedures and in decisions, particularly those that potentially affect the communities in which they live and work. To this end the ECE *Guideline on Access to Environmental Information and Public Participation in Environmental Decision-making* (UN ECE 1996b) could be applied, and soon the convention on access to environmental information and public participation on environmental decision-making which is under preparation and will be signed at the next Ministerial Meeting in Denmark in 1998.

14.5
Conclusive Remarks

The coastal zone is the major attraction pole for settlement and economic development. It has also a high potential for recreation and tourism. The attraction of resources and opportunities has led to increased inhabitation, stresses, deterioration conflicts between users, and between exploitation and conservation. Planning strategies must be based on detailed area knowledge, mapping zoning, analyses, evaluations and inventory taking. A coordinated policy of research, planning and management backed by public support will foster positive action.

Required actions cover preservation and protection of fragile environments through legislation, purchase control, promotion of non-destructive uses, elimination of waste disposal operations, production of descriptive, diagnostic and prescriptive maps, pollution abatement, building regulation, biomass exploitation, minimization of natural hazards, preservation of high quality soils, EIA, development of programmes of public information and education, and nurturing of general awareness of the need to protect the coastal zone and promote conservation and management co-operation (Charlier and Charlier 1995).

The casual approach to coastal zone occupation and use must bow to the imperatives of the coming decades: scientific approach and careful planning, resistance to selfish private interests, a "global" approach, and synchronization of legislation. In this regard, the transboundary conventions have a useful role to play. A coherent integrated policy of inventory, diagnosis planning and management and research is needed.

The existing regional legally binding instruments and "soft-law" recommendations constitute a solid framework for sustainable integrated coastal area management. It is now mainly a question of elaboration and implementation at the national and transboundary levels – both bilateral and multilateral – to put the legal and institutional framework into practice.

In order to integrate environmental and coastal area development decision-making processes in a more comprehensive way, it is necessary not only to elaborate and implement different environmental policy sub-sectors, but also to apply a common glue for these broad policy fields. The polluter-pays principle, precautionary principle, and the principle of preventing environmental damage form one component of this glue, spatial and land-use planning, as well as economic instruments, the second one (ibidem).

References

Charlier C, Charlier R (1995) Sustainable multiple-use and management of the coastal zone in environmental management and health. Vol 6, Issue 1, ISSN 0956-6163

European Commission (1996) Brussels European sustainable cities. Eurepean Commission Directorate General XI, Expert Group on the Urban Environment, Brussels

European Environment Agency (1995) Europe's environment. The Dobris Assessment EEA, Copenhagen

United Nations Economic Commission for Europe (1993) The Rio follow-up at regional level. A review of major regional policy implications of the outcome of the United Nations conference on environment and development. United Nations, Geneva

United Nations Economic Commission for Europe (1994) Environmental conventions elaborated under the auspices of the UN/ECE. United Nations, New York and Geneva

United Nations Economic Commission for Europe (1995a) Environmental Programme for Europe. Working group of senior governmental officials, endorsed by the ministerial conference Environment for Europe. Sofia, Bulgaria

United Nations Economic Commission for Europe (1995b) Water series No. 2. Protection and sustainable use of waters. Recommendations to ECE Governments. United Nations, New York and Geneva

United Nations Economic Commission for Europe (1996b) Guidelines on sustainable human settlements planning and management. United Nations, New York and Geneva. ECE/HBP/95)

United Nations Economic Commission for Europe (1996b) Guidelines on access to environmental information and public participation in environmental decision-making. United Nations, New York and Geneva

United Nations Environment Programme Nairobi (1982) Convention for the protection of the Mediterranean Sea against pollution (Barcelona Convention). United Nations, New York

van der Wal (1993) Sustainable urban development: Research and experiments. Proceedings of the PRO/ ECE workshop in Dordrecht, November 1993. Delft University Press

CHAPTER 15

The Europipe Landfall Project

Henning Grann

15.1
Introduction

This chapter provides an overview of the natural gas pipeline network in the North Sea and the development of the European natural gas market. In addition, a more detailed description is given of the Europipe landfall project including the project planning and construction. Over the next few years a major offshore natural gas pipeline transportation system will be completed on the Norwegian continental shelf. From the gas fields, thousands of kilometers of large-diameter pipelines bring the natural gas to the European continent. The Europipe transportation system is a part of this integrated pipeline network designed to bring large amounts of Norwegian gas to several countries in Europe. The Europipe Transport System is owned by a joint venture including the companies Statoil, Norsk Hydro, Saga Petroleum, Esso, Shell, Total, Elf, and Conoco. Statoil is the operator of this joint venture.

15.2
Background

In the early 1960s the consumption of natural gas in OECD Europe was only 2% of the total energy demand. However, the discovery of the large Groningen gas field in The Netherlands was the first step in a considerable change in this picture. In 1964 Algeria started export of liquified natural gas to the UK and Spain and in 1968 the former Soviet Union began their export of gas to Western Europe.

In the early 1970s a consortium of continental gas buyers purchased significant volumes of gas from the Norwegian offshore Ekofisk field with the objective of securing a long-term supply of gas to Western Europe. As a result the Norpipe gas pipeline to Emden in Germany was built and started up in 1977. Following the signing of the Troll gas sales agreement in 1986, it was decided to develop the Norwegian Troll and Sleipner fields. Consequently, the Zeepipe pipeline to Zeebrugge in Belgium was built and commissioned in 1993.

By this time the natural gas share of the energy consumption in the European OECD countries had increased to 17%, the total gas demand was about 320 billion m^3 year^{-1} and a continuing increase was forecasted. As of now, the total Norwegian contracted gas volumes have reached a level of 62 billion m^3 year^{-1}. The increasing emphasis on natural gas is the result of European governments' strategy to improve the security and flexibility of their energy supplies. Moreover, substitution of coal and oil by natural gas is resulting in substantial environmental improvements and has become an important instrument for European governments in their attempts to realize their objectives of reduction and stabilization of CO_2 emissions, which are viewed as a significant contribution to the greenhouse effect and the resulting global warming.

Already in the mid 1980s it became apparent that there would be a lack of gas transportation capacity from Norway to the European continent in the foreseeable future due to higher sales volumes under existing agreements and due to new additional gas sales agreements signed with German and Dutch buyers. In order to meet the increased gas supply obligations a number of supply alternatives were examined including:

- Upgrading of existing pipeline systems to higher operating pressures
- Change of gas delivery point for certain requirements
- New offshore pipeline to Denmark with an onshore pipeline through Denmark and Germany to Emden
- New offshore pipeline to The Netherlands or to Germany with onshore pipeline to Emden

In the autumn of 1991 it was decided to develop the documentation required for permission to land Europipe, a new offshore pipeline, in Germany with an onshore connection to Emden. Presently two more pipelines are under planning and development, namely Europipe II (parallel to Europipe I) and NORFRA to be landed on the northern coast of France near Dunkirk, bringing the total number of gas pipelines from Norway to the European continent to five.

15.3
Project Planning and Approval

Initial contacts between Statoil and the German authorities were made in 1985 based on the first feasibility study of a pipeline landfall in Germany. This was followed by a preliminary evaluation of some ten alternative routes along the German coast in early 1990. In 1986 the Wadden Sea National Park of Lower Saxony was formally established covering a major part of the North Sea coastline of Lower Saxony.

In accordance with the Regional Planning Procedure, Statoil submitted its first formal request for planning permission early in 1991 with pipeline landfall via the island of Norderney, which was the alternative originally favored by the German authorities. By this time the political situation in Lower Saxony had changed and led to the formation of a coalition government between the Social Democrats and the Green Party. As a result, environmental issues were given considerable attention on the political agenda and the environmental organizations were gaining an increasing influence in the decision process. This made it increasingly difficult to develop a pipeline project which would satisfy the interests of the non-governmental environmental organizations, the authorities and the public in general. Consequently, numerous alternative pipeline routes, technical solutions and associated environmental impact assessments were developed for review by the authorities.

After long, complicated and at times quite frustrating discussions, hearings and negotiations, Statoil was finally in November of 1992 granted planning permission to land Europipe through the Accumer Ee tidal inlet and a subsurface tunnel under the tidal flats as the most costly but environmentally most favorable project alternative. After further optimization of the selected alternative, final construction permission was given on 27 October 1993.

Having been through this difficult planning permission process, it seemed logical to anticipate the need for future additional pipeline capacity by utilizing the spare space in the pipeline trench and the tunnel. Therefore, it was decided to apply for permission to lay 12 km of pipeline with tie-in connections in both ends parallel to Europipe through the landfall area of the National Park, thereby avoiding future construction work in this environmentally sensitive area. This permission was given in early 1994.

15.4
Europipe Project Facilities

The total Europipe project can be subdivided into:

1. Off-shore facilities consisting of: a) A new riser platform, bridge – connected to an existing platform in the Sleipner field in the Norwegian sector of the North Sea b) A 630-km, 40-in.-diameter pipeline from the new riser platform to the German North Sea 3-mile zone.
2. Landfall facilities and/or construction activities consisting of:
    a   Dredging of a 12-km, 2-m-deep, 12-m-wide trench from the 3-mile zone through the Accumer Ee tidal inlet
    b   Laying of two parallel 40-in. pipelines in the trench
    c   A 2.6-km, 3.8-m-diameter concrete tunnel through which the pipeline was brought on-shore under the tidal flats and the mainland dike
    d   A 12-m-diameter, 20-m-high cylindrical tie-in chamber, connected to the tunnel, and enabling recovery of the tunnel boring machine and installation of the tie-in connections between the land and the off-shore pipeline
    e   Laying of two parallel 40-in. pipelines through the tunnel
    f   Backfilling of the dredged material.
3. On-shore facilities consisting of:
    a   A 50-km, 42-in. diameter on-shore pipeline
    b   A gas receiving station at Dornum, some 5 km from the tunnel
    c   A gas metering station at the contractual delivery point in Emden.

15.5
Europipe Landfall Facilities/Activities

15.5.1
Dredging

The pipeline landfall section is about 12 km long, extending from the 3-mile zone outside the islands of Baltrum and Langeoog, crossing a sandbank, following the Accumer Ee tidal inlet between the islands, continuing into the shallow Accumersieler Balje before reaching the tidal flats. The pipeline route crosses a dynamic morphological area, where the tidal channel is found to shift sideways with a cycle time of about 45 years. Hence, a morphological design basis had to be established, taking the dynamics into account, as well as observed long-term rise of the water level in the area. The design basis chosen for the burying of the pipeline was an exposure probability of the top of the pipeline of 5% in the lifetime, i.e. once in 1000 years. This implied that the minimum cover over the pipeline was to be 2 m. For the purpose of burying the pipeline permanently under the sea floor, a trench of varying depth and with a trench bottom width of 12 m had to be dredged. In addition, a 100-m-wide vessel access channel had to be cut through the sandbank. In total some 3.4 million m^3 of material containing sand, clay and peat was removed to be used for beach nourishment, disposal on-shore, or to be brought to an interim storage area off-shore for later backfilling, dependent on the type of the dredged material and its environmental impact.

Depending on water depth and local conditions, the pipeline route was divided in sections, each requiring different construction methods and equipment.

The near-shore section is 3 km long with water depths varying between 9 and 18 m. This area constitutes a natural shelf in front of the sandbank and is exposed to severe wave conditions. In this section, large seagoing trailer suction hopper dredgers were used. The material was loosened mechanically or by waterjets, hydraulically extracted and loaded into the hopper. In areas where peat-containing material was expected, a bucket dredger was used in order to avoid peat from being set adrift, contaminating the beaches or, even worse, causing problems for the local fishermen. Peat-containing material was loaded in barges and brought on-shore in Emden 100 km away.

The sandbank section has a typical profile with depths less than 9 m, in some areas only 3 m, forming a shallow water barrier at the ocean side of the islands. This section is about 2.8 km long and is characterized by the prevailing presence of breaking waves and by tide-dependent restrictions on marine traffic. In the sandbank section, a cutter suction dredger was used. The cutter suction dredger loosens the material by means of a cutting head and conveys it hydraulically to seagoing barges for transport out of the area. This dredging method is particularly effective in areas where wide trenches are to be dredged. In addition, trailer suction hopper dredgers were used in parts of the sandbank and for maintenance dredging. The tidal channel is about 4 km long and is the most narrow passage between the islands, with a water depth of up to 22 m and high current velocities which require special construction methods and equipment. In the tidal channel various size trailer suction hopper dredgers were used. A large stiff clay ridge in the tidal channel was dredged using a cutter suction dredger.

The Accumersieler Balje section is 2 km long and is the inner shallow part of the landfall with water depths of 2–5 m and strong tidal currents. In the Balje section two spud pontoon dredgers equipped with backhoes were employed. This method was chosen due to the shallow water depth and large inclusions of peat in the dredged material. All peat-containing material was transported to Emden and used in the construction of an artificial biotop, which Statoil made as an environmental compensation measure. About 400 000 m^3 of clay material was disposed of 60 km off-shore on a permanent dumping site. All sandy material was brought out of the National Park area to an interim off-shore storage area for later use as pipeline trench backfilling material.

15.5.2
Pipelaying in the Tidal Channel

The off-shore pipelaying was performed by large pipelaying barges with a capacity to lay up to 3.5 km of large diameter pipe per day, virtually independently of seasonal

Fig. 15.1. Castoro Sei off-shore pipelaying barge

weather conditions. In the near-shore, sandbank and tidal channel areas, however, smaller pipelaying vessels had to be used. In addition, 34 piles of 40-in. diameter were installed and used as anchors during the pipelaying operation. The first 10-km pipeline in the landfall area was laid in 45 days (see Fig. 15.1).

For the last 2 km in the inner shallow part of the Accumersieler Balje, a pipe hauling method was originally envisaged. This was later changed to the use of a small shallow draught pipelaying barge which had been modified to enable the handling and laying of large diameter pipe. With its 58-m length, 11-m width and 2-m draught, this is probably the smallest barge ever used for the laying of large diameter pipelines. The vessel did not have the capacity to carry spare pipes while laying and with its limited lifting capacity, pipeline pick-up operations did require much attention and special precautions. Despite its limitations, this small modified vessel proved to be a success, laying up to a maximum of 21 joints a day.

15.5.3
The Tunnel

For the construction of the tunnel under the mudflats a special boring machine was developed and purpose built to drill the total length of more than 2500 m without breakdown or replacement of parts (see Fig. 15.2). Furthermore, the German authorities only allowed human intervention in the tunnel during weekends, when no drilling operation took place, and work was limited to preventive maintenance only. For these reasons the facilities were designed and built for remote control and operation of the boring machine, the Bentonite lubrication system and all utilities from a control center in

Fig. 15.2. Tunnel boring machine

the tunnel entrance shaft. The tunnel itself was constructed of a total of 638 concrete sections, each 4 m in length, 3.8 m outside diameter and with a wall thickness of 0.4 m. For every 4 m of drilling, the machine was stopped and a very powerful jacking system pushed a concrete section into the tunnel opening (see Fig. 15.3). This operation was repeated for the entire length of the tunnel. To facilitate the jacking operation, injection points for continuous Bentonite lubrication were provided for each 100 m of tunnel length.

The tunnel boring machine was ready for operation in February 1994. The planned average drilling speed was estimated at 15 m per day with expected tunnel completion in October 1994. However, the introduction of an extra shift improved the schedule to an expected completion in August. The soil conditions were anticipated to be favorable for this type of drilling; however, detailed soil data were limited due to restrictions on soil sampling in the mudflat area. As a result, the extent and structure of sand, clay and peat formations had to be assessed on the basis of rather insufficient information. The concern in this connection was that the lower part of the tunnel would reach into glacial sand formations, in which boulders of varying sizes are known to be present. The boring machine was capable of handling boulders up to 35 cm in diameter; however, if larger boulders were encountered, this could necessitate cumbersome and time-consuming intervention from the surface.

After a few initial technical problems the tunnel operations turned out to be a success. The average drilling rate exceeded the estimate by a factor of almost 2, no large boulders were found, and the drilling operation was completed in exactly 100 days. This probably constitutes a new world record for this type of tunnel construction in terms of drilling rate and total length of tunnel as well.

Fig. 15.3. Tunnel concrete section jacking system

15.5.4
The Tie-in Chamber

The physical connection of the off-shore pipeline to the land pipeline turned out to be a real technical challenge, particularly when considering all the restrictions imposed on the work in the National Park. After rejection of several alternatives, the concept of a sub-sea floor tie-in chamber was developed. The chamber itself was a 12-m-diameter 20-m-high steel cylinder with an 8-m-high removable steel funnel on top, enabling the pipeline tie-in connections to be made inside the chamber under dry, atmospheric conditions, and enabling recovery of the boring machine in a dry operation, rather than having to transport it back through the full length of the tunnel. For the tranportation and installation of the tie-in chamber, a special catamaran-type vessel was purpose built (see Fig. 15.4).

The late decision to lay a 12-km parallel pipeline through the landfall area had implications for the construction of the tie-in chamber in terms of additional stress imposed on the chamber. As a result, additional piling was required, causing some delays. The situation was further complicated by the fact that the tunnel boring machine was progressing at higher than estimated speed. An interruption of the tunnel drilling was to be avoided, as sediments around the tunnel sections might settle, making restart of the jacking operation uncertain. In the end, the scheduling problems were resolved, and in early June, champagne could be served in the tie-in chamber as the boring machine penetrated the chamber with a deviation from the target of just 11 cm (see Fig. 15.5).

Fig. 15.4. Tie-in chamber installation site and catamaran vessel

15.5.5
Pipelaying in the Tunnel

After recovery and dismantling of the boring machine and the tunnel jacking facilities, a welding station was set up on-shore in front of the tunnel entrance shaft and the two approximately 2500-m lengths of 40-in. parallel pipeline were gradually welded and hauled through the tunnel from the tie-in chamber (see Fig. 15.6). Upon completion of this job, the tie-in between the two off-shore and on-shore pipelines was made, the tie-in chamber was cut to a level of 2 m below the seabed and both the tunnel and the remaining part of the tie-in chamber were filled with a special temperature-stable concrete mixture.

15.5.6
Backfilling of the Dredged Material

After completion of the pipelaying and the tie-in operations in the tie-in chamber, the trench and the access channel were backfilled with sandy material from the interim storage area. Backfilling took place by means of trailer suction hopper dredgers of different sizes, depending on available water depth. The material dumping was carried out in a controlled manner using discharge pipes to avoid dispersion and loss of material in the strong tidal currents. The total amount of sand being backfilled was approximately equivalent to the amount of material being dredged; thus, the mass balance in the system was re-established, which was one of the conditions given in the authority permission.

Fig. 15.5. The tunnel boring
machine penetrates the tie-in
chamber

Fig. 15.6. Pipelines being
hauled into the tunnel

15.6
Environmental Management of Construction Activities

In order to meet the stringent German environmental requirements and to ensure a high level of environmental performance, a comprehensive environmental protection plan was developed for each of the project planning, construction and operation phases. This plan included environmental studies and impact assessments, ecological monitoring programs, environmental audits and, as a special feature, a detailed program for environmental management of the construction activities. For the latter, the British Standard BS 7750 was used as a guide in the development of a methodology which enabled the identification of environmentally critical operations, development of effective protection measures, and, above all, development of simple work instructions for the first line construction supervisors and their crews. For more details see Schuchardt and Grann (this Vol.).

Considerable training and instruction efforts were required to effectively implement the environmental management system and this was initially viewed by the contractors with much scepticism. However, as the project progressed, the attitudes changed positively and in the end some contractors even expressed the view that they had gained a valuable experience which would benefit their competitiveness in future construction projects. There is no doubt that the environmental management approach adopted has been effective in preventing the occurrence of environmental incidents.

15.7
Concluding Remarks

The Europipe landfall project went through a very difficult planning and approval process. The development of a project concept which would satisfy all interested parties proved to be a considerable challenge to the German authorities as well as to Statoil. New and unproven technology was needed to comply with the strict authority requirements for working in the National Park area. With this background, however, the project was technically well executed, considerable care and attention was given to the environmental aspects of the project, and it is believed that the ecological monitoring programs will document that the Europipe project has been implemented without permanent adverse effects on the National Park ecosystem.

As regards the early project planning and development phases, important lessons have been learned, particularly with respect to the development of an improved integrated management approach which incorporates the environmental impact assessments throughout the entire project. For more details on this point see Schuchardt and Grann (this Vol.). This experience is being utilized in the Europipe II project, presently under construction.

The total cost of the Europipe project was in the order of DM 4 billion, of which DM 1 billion was spent on the landfall facilities. The additional costs of the tunnel solution under the sensitive mud flats was DM 250 million as compared to the original plan of a direct open trench via the island of Norderney. The tunnel alternative was chosen by the authorities after a long planning and approval process as the environmentally best solution. From a total environmental point of view this decision has been questioned, considering the resource- demanding tunnel construction works and the extensive additional dredging works with their temporary adverse ecological impact. In retrospect it is being debated whether the high additional project costs were in fact environmentally justified or merely the result of political pressures and considerations.

References

RSK Environment Limited, Environmental Report of the Europipe Project
Schuchardt B, Grann H (this Vol.) Towards an integrated approach in environmental planning: the Europipe experience
Statoil, Various internal project reports

Part V
Environmental Science

Embankments and Their Ecological Impacts: A Case Study from the Tropical Low-Lying Coastal Plains of the Deltaic Sunderbans, India

Asokkumar Bhattacharya

16.1
Introduction

The deltaic Sunderbans occupy 4267 km² along the northeast coast of India (21°32′–22°40′ N′; 88°05′–89°0′ E), and are bounded by the Hooghly river in the west, the Haribhanga river in the east, the Bay of Bengal in the south, and the Dampier-Hodges demarcation line in the north (Fig. 16.1). There are 54 islands of various shapes and sizes separated by a network of tidal channels, inlets and creeks, some of which act as pathways for both upland freshwater discharge and the to-and-fro movements of flood and ebb waters (Mandal and Ghosh 1989). In many areas the island margins are protected by embankments whose combined length amounts to about 3500 km (Sharma 1994). Some of these large-scale constructions have sluices for the drainage of trapped saltwater.

Embankments act as major elements in the coastal defence of the low-lying coastal plains and islands. They are constructed near the intertidal borders of rivers, tidal inlets and creeks in order to protect agricultural and inhabitable land from saltwater inundation. Their combined length of roughly 3500 km is often strongly reduced as a result of breaching during devastating seasonal floods.

The embankments are maintained by the state government in order to protect the interests of local inhabitants in the Sunderbans region. In a multi-faceted attempt to integrate such activities within a broader framework, the present study deals with the function, design and efficiency of these large-scale constructions as well as the economic consequences of breaching and ecological impacts of embankments in the Sunderbans coastal plains.

16.2
Physical Setting

The present-day configuration of the low-lying Sunderbans coastal plains of northeast India is the result of the combined effects of deltaic depositional processes in the Ganges and Brahmaputra rivers, estuarine depositional processes, tidal depositional processes in the Bay of Bengal, tectonic subsidence of the Bengal basin during the Tertiary, and a general southerly tilting of the basin during recent times (Biswas 1963; Morgan 1965 Sengupta 1972). The region shows a combination of land and water bodies incorporating 1) tidal creeks and inlets (2350 km²), 2) halophytes and mangroves (1750 km²), and 3) recently raised shoals (217 km²); (Naskar 1996).

The coastal plains slope gently seawards, and lie largely 7–8 m above mean sea level. The tidal regime is meso-macrotidal and semi-diurnal (Davies 1964; Bhattacharya 1993).

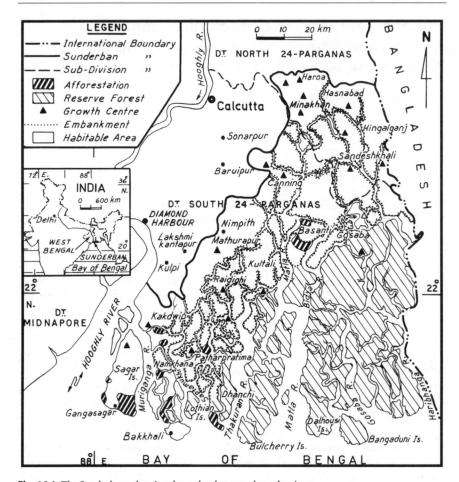

Fig. 16.1. The Sunderbans showing the embankments along the rivers

The mean tidal range at spring tide is 4.87 m, and 2.13 m at neap tide. It can measure >6 m under extreme meteorological conditions. Flood and ebb current velocities vary in the range 1.5–3.8 m s^{-1}. The tidal length is 60–80 km, a value which increases to 290 km in the Hooghly river.

Salinity varies from 8‰ in the monsoon season (August–September) to 20‰ in the pre-monsoon season (March–April). Southwesterly and northeasterly winds dominate during the SW and NE monsoon events, respectively (Banerjee 1972). The maximum wind velocity is 16.7–50 km h^{-1} in April–June, decreasing to minimum values of 10.7–11.8 km h^{-1} in December–February.

The Bay of Bengal is cyclone-prone, and can experience three to four such events a year during which time maximum wind velocities can reach 80–140 km h^{-1}. Cyclones initiate large-scale littoral drift, severe coastal erosion, and a substantial accumulation of sediments in nearshore waters. In addition, cyclonic activity causes a heavy loss of life and property, particularly when strong wave events follow in close succession.

Wave heights can exceed 2.5 m during cyclones, and wave periods are >14 s (Banerjee 1972). In contrast, wave heights measure 0–0.6 m, and wave periods are 5–7 secs during the calm winter season. These values increase to 1.8–2.4 m and 12–14 s, respectively, during the rough summer months. The average annual rainfall is 1750–1800 mm, 80% of which occurs from May to September. Average monthly air temperatures vary between 34–36° C in April–June and 21–25° C in December–January. Relative humidity is >80% in June–September, and >75% in October–May.

The major economic activities in the Sunderbans region involve fishing, forestry, industry, navigation, urbanisation, agriculture and recreation, many of these being threatened by erosion in the coastal zone. Erosion can occur as either a gradual denudation or a catastrophic event. Gradual erosion reflects the effects of 1) strong tidal currents at times of high tidal ranges (>6 m), and 2) a moderate to strong littoral drift. Catastrophic erosion events are linked with 1) cyclonic storms and hurricanes, 2) tidal bores, and 3) tectonic and neotectonic movements (Bhattacharya and Das 1994; Bhattacharya 1997a).

16.3
Embankment Function and Design

In the Sunderbans region embankments mainly serve to:

- Prevent the erosion of river banks
- Protect the hinterlands from saltwater, particularly the agricultural and inhabited areas situated immediately above the intertidal zone
- Control water runoff in tidal channels for flood alleviation
- Guard against pirates as well as wild animals such as the Royal Bengal tiger
- Protect low-lying areas from tropical cyclones.

Embankments are large-scale constructions commonly made of bricks, concrete, boulders (mostly laterites) and mud with or without revetments and rip raps. Facing the sea or rivers they are often reinforced at their bases by boulder mounds which thereby provide so-called toe protection. Four main types of embankments are described below.

16.3.1
Boulder Embankments

The most common construction design of boulder embankment makes use of laterite boulders from the Middnapore district. The rough surfaces of these boulders impart additional safety by effectively dissipating wave energy (Petts and Calow 1996). Banks of earth about 15 m wide are constructed along the seashore and riverbanks (Fig. 16.2). Facing the river the earthworks have slopes measuring 25–30 degrees. They are subsequently covered by 8 to 10-cm-thick filter beds of crushed brick bat overlain by regularly positioned laterite boulders. The boulder layers are a 75 cm thick. The constructions are 4–5 m high in all.

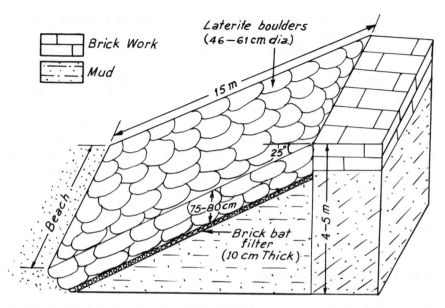

Fig. 16.2. Boulder embankment in Beguakhali, Sagar Island, showing inner core of earth with filter bed covered by laterite boulders

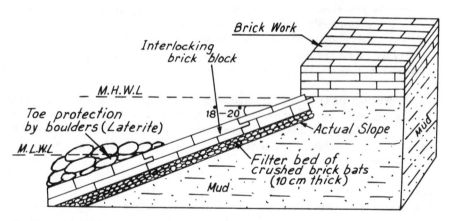

Fig. 16.3. Brick-block revetment at Chemagari creek margin, Sagar Island

16.3.2
Brick-Block Revetments

Brick-block revetments are large-scale constructions which often serve to protect creeks and riverbanks from erosion by currents and waves. Interlocking brick blocks overlaying filter beds of crushed brick bat (10–12 cm thick) are used for this purpose (Fig. 16.3). In addition, laterite boulders are commonly employed for toe protection. The slopes facing the rivers generally measure 18–20 degrees.

16.3.3
Timber and Bamboo Pile Bulkheads

Timber and bamboo pile bulkheads constructions serve to retain sediment fill on steeply sloping (30–35 degree) riverbanks (Fig. 16.4). The heights of the bulkheads vary in the range 2.5–3 m. Toe protection is generally provided by boulders or gunny bags filled with sand or mud.

16.3.4
Gunny-Bag Embankments

The stacking of gunny bags filled with sand, mud or coal dust often serves as a temporary remedial measure along damaged sections of embankments.

16.4
Engineering Performance of Embankments

Heavier rock or concrete embankments are built in the more exposed macrotidal settings of large estuaries, rivers and tidal inlets (e.g. in the Hooghly, Saptamukhi, Thakuran and Matla rivers; Fig. 16.1). In contrast, earth banks stabilised by mangrove vegetation seem to be effective enough under low-energy mesotidal and microtidal conditions further upstream (Bhattacharya 1997b). The main reasons for embankments sometimes failing to perform satisfactorily as protective large-scale constructions include improper design and construction, natural destabilisation, and inadequate maintenance.

16.4.1
Improper Design and Construction

Improper design and mechanical defects arising from uncompleted building activities are generally the main causes of embankment failure. Because these constructions are

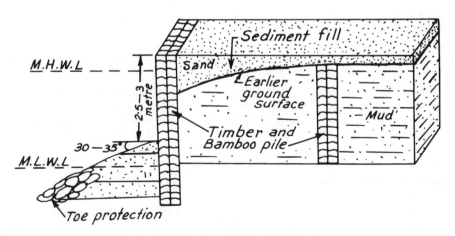

Fig. 16.4. Timber and bamboo pile bulkhead at Matla river bank, Canning

often too low, their inefficiency results largely from improper design. Thus, frequent and large-scale overflowing during storm surges and tidal bores is probably the single most important consequence of the improper design of embankments in the Sunderbans region.

Mechanical defects resulting from the stalling of construction schedules can also pose large problems. After the completion of the earthworks, construction activities are often interrupted, sometimes for several years, because of a paucity of funds or even plain negligence. Without protective layers of boulders or brick blocks, mud banks in particular develop desiccation cracks whereby leakage can result in the disruption of the uncompleted constructions at later stages (Fig. 16.5a). In the Bay of Bengal simple tunnels/channels and sluices are widely used for sewage disposal and the drainage of saltwater from inland ponds. Because these passages are often inadequately fortified, tidal scouring can severely weaken and ultimately even destroy the embankments in the vicinity.

16.4.2
Natural Destabilisation

Growth of the roots and pneumatophores of mangroves also decreases the resistance and durability of embankments by creating interconnected openings and passages through which leakage occurs (Fig. 16.5b). The burrowing activities of the mud lobster or ghost shrimp *Thalassina anomala* have similar negative effects on bank stability. Indeed, these organisms excavate passages as much as 2 m deep in the upper intertidal and supratidal zones of rivers (Mandal and Ghosh 1989).

The effects of various factors reducing the stability and, therefore, the performance of embankments are severely compounded by the liquefaction of the subsoil below the water-table. Liquefaction can lead to 1) the tilting of the embankments (Fig. 16.5c), 2) the cracking of mud below the embankments (Fig. 16.5d), and 3) the lateral spreading of the subsoil along gently sloping riverbanks (Fig. 16.5e); Youd and Keefer 1981; Youd 1984; Johnson and de Graff 1988).

16.4.3
Inadequate Maintenance

Last but not least, the failure of embankments to provide effective protection is also explained by the oft tardy and piecemeal nature of repair operations. These can generally be carried out only after the monsoon season because persistently hostile weather conditions render repair work largely impossible before. Rapid repair work can be severely impaired also by logistic problems in procuring boulders for construction from distant regions.

16.4.4
Embankment Breaching

During severe cyclonic storms strong wave action can cause the breaching of embankments in the Sunderbans region. Consequently, tidal waters flood the low-lying villages and adjacent farmland, causing large-scale devastation on the coastal plains. The vil-

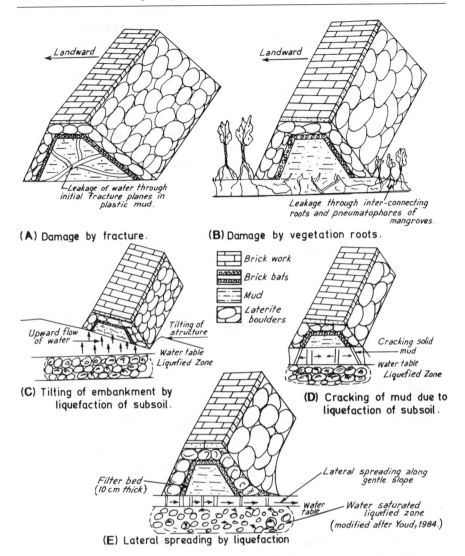

Fig. 16.5. Schematic representation of damages suffered by embankments. **a** Damage by fracture. **b** Damage by roots. **c** Damage by liquefaction resulting in tilting. **d** Damage by liquefaction resulting in cracking of mud base. **e** Damage by liquefaction resulting in lateral spreading. (Modified from Youd 1984)

lages of Patherpratima, Kakdwip and Namkhana as well as the islands of Sagar, Gosaba and Basanti (Fig. 16.1) are particularly badly affected by floods almost every year. In 1997 some 5000 people were marooned in the fringe areas of the Sunderbans when a cyclone hit the coast, causing 6.35-m-high waves to breach the embankments on the 25 August that year.

Wave-induced damage to embankments can be extensive, involving distances varying from one to several kilometres. The force of the water causes the displacement of the revetment boulders and concrete slabs at the foot of the embankments, thereby exposing the mud earthworks at the core of the constructions. Breaches tens of metres long can occur repeatedly along the sandy beaches and muddy tidal flats.

Along the margins of the embankments strong wave action occasionally causes the formation of to 3–4-m-deep ditches whose surface areas can reach 10 m² in each case. These ditches promote the undercutting of local sediments, thereby facilitating the collapse of embankments during subsequent floods. The catastrophic breaching of the embankments and the ensuing devastation in the hinterlands incur enormous economic costs in the Sunderbans region. In addition, much man-power and money is spent every year on the maintenance and repair of these large-scale protective constructions.

16.4.5
Ecological Impacts of Embankments

In combination with other large-scale constructions, man-made embankments cause both short-term and long-term perturbations of the ecological equilibrium of the Sunderbans deltaic plains. These negative effects can be grouped into three main categories.

1. As large-scale constructions embankments substantially hinder natural siltation in floodplains. Consequently, excessive siltation has caused the marked shallowing of rivers, inlets and creeks. The shallowing of almost all rivers in the Sunderbans region has currently reached alarming proportions. This state of affairs is particularly noticeable in the rivers Ganges, Saptamukhi and Thakuran (Fig. 16.1). In turn, the shallowing of rivers enhances the sinuosity and migration of downstream tributaries and channels.
2. The migration of waterways and frequent flooding occur mainly during periods of high precipitation in the July–September monsoon period. It has recently been documented that the tidal network and drainage systems have increased in length in the region, a trend accentuated by intermittent tectonic subsidence of the Bengal Basin. This is reflected in the intricately braided patterns of the rivers on the Sunderbans coastal plains. The breaching of the embankments during the monsoon season can greatly damage reclaimed arable land which then reverts to being marshland (Chatterjee 1972). Thus, temporal and spatial shifts in the flora and fauna reflect differing sets of pre-embankment and post-embankment environmental conditions.
3. The man-induced loss of valuable mangrove wetlands is a direct result of embankment construction policy. These large-scale constructions often cause the abrupt truncation of the wetlands because they are situated in intertidal habitats favoured by mangrove forests. Invasion by marsh vegetation and land reclamation then hinder the survival of mangroves landwards (Fig. 16.6).

It has been demonstrated that the loss of mangrove ecosystems imparts a serious ecological imbalance in tropical coastal regions (Woodroffe 1990). Most importantly in the present context, the buffering capacity of the mangroves is seriously impaired (e.g. Mazda et al. 1997). The hydrodynamic forces acting on the embankments are thus stronger than they would be if these constructions were built further inland beyond an intact mangrove belt.

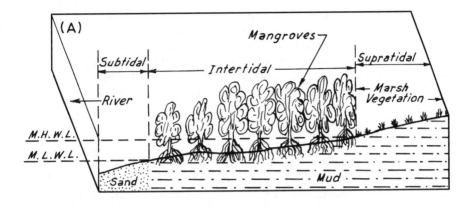

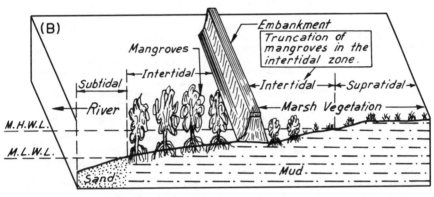

Fig. 16.6. Truncation of mangroves by embankments constructed in the intertidal zone. A Entire expanse of the intertidal zone is occupied by mangroves. B Mangroves are truncated by embankments in the intertidal zone. Landwards the intertidal area behind the embankments is occupied by marsh vegetation

16.5
Conclusions

The qualified regulation of embankment height is an important aspect in the decision-making process of maximising the efficiency of protective measures such as embankments on the one hand, and minimising construction costs on the other hand. In order to come to a meaningful solution in the matter, the long-term monitoring of catastrophic high water stands is necessary.

At present there exists evidence of shoreline erosion in some regions of the Sunderbans coastal plains (Bhattacharya 1997a). This indicates a slow and irregular rise in sea level which would tend to displace the mangrove forests landwards. These natural adjustment processes should be integrated into long-term policy-regulation of embankment construction. Similar considerations have long been voiced in regulating the conservation of other animal-plant communities, including those of tidal marshes (Daiber 1986).

It is suggested that, in future, the construction of new embankments should be restricted to areas above the supratidal zone in order to provide sufficient space for the landward growth of the mangroves. Safeguarding these forests would also impart greater stability and durability to the embankments. Because of the huge construction and maintenance costs involved, these considerations would be of key importance in any cost-benefit analyses of embankment construction in the Sunderbans coastal plains.

References

Banerjee BN (1972) Navigation in a tidal river with particular reference to river Hooghly. In: Bagchi KG (ed) The Bhagirathi-Hooghly basin. Proc Interdisc Symp. Calcutta Univ Publ, India, pp 154–167
Bhattacharya A (1993) Backwash-and-swash-oriented current crescents: indicators of beach slope, current direction and environment. Sediment Geol 84:139–148
Bhattacharya A (1997a) Morphodynamics of the coastal tract of West Bengal, northeast India, and evidence of shoreline displacement. Int Sem Quaternary sea-level variation, shoreline displacement and coastal environment. Tamil Univ, Thanjavur, 20–26 Jan 1997. Abstr Vol, pp 20–21
Bhattacharya A (1997b) Hazards related to embankments behind the intertidal zone: a case study from the tropical low-lying coastal plains of the deltaic Sunderbans, eastern India. In: Vollmer M. Delafontaine MT (eds) Proc 1ˢᵗ Int Symp Large-Scale Constructions in Coastal Environments. 21–25 April 1997, Norderney (Germany). Ber. Forschungszentrum Terramare 1:30–31
Bhattacharya A, Das GK (1994) Piecemeal mechanism of bank erosion following subsidence: a case study from Thakuran river of deltaic Sunderbans, West Bengal. J Ind Soc Coast Agric Res 1(2):231–234
Biswas B (1963) Results of exploration for petroleum in the western part of the Bengal basin, India. Proc 2nd Symp Dev Petrol Res ECAFE. Min Res Dev Ser 18:241–250
Chatterjee SP (1972) The Bhagirathi-Hooghly Basin. In: Bagchi KG (ed) The Bhagirathi-Hooghly basin. Proc Interdisc Symp. Calcutta Univ Publ, India, pp 19–24
Daiber FC (1986) Conservation of tidal marshes. Van Nostrand Reinhold Company, New York
Davies JL (1964) A morphogenic approach of world shorelines. Z Geomorphol 8:27–42
Johnson RB, Graff JV de (1988) Principles of engineering geology. John Wiley & Sons, New York
Mandal AK, Ghosh RK (1989) Sunderban – a socio-ecological study. Bookland Priv Ltd, Calcutta
Mazda Y, Magi M, Kogo M, Hong PN (1997) Mangroves as a coastal protection from waves in the Tong King delta, Vietnam. Mangroves Salt Marshes 1(2):127–135
Morgan JP (1965) Depositional processes and products in the deltaic environment. In: Morgan JP (ed) Deltaic sedimentation. Soc Econ Pal Min Spec Publ 16. Blackwell, oxford, pp 30–47
Naskar KR (1996) Sunderbaner Mohonaban O tar Sankat. In: Bengal activists for biodiversity and environment. Banglar ban O Kalkatar Paribahan, pp 11–16 (in Bengali)
Petts G, Calow P (eds) (1996) River restoration. Blackwell Science, London
Sengupta S (1972) Geological framework of the Bhagirathi-Hooghly basin. In: Bagchi KG (ed) The Bhagirathi-Hooghly basin. Proc Interdisc Symp. Calcutta Univ Publ, India, pp 3–8
Sharma BC (1994) Lower Sunderban. The Asiatic Soc Publ, India
Youd TL (1984) Geologic effects – liquefaction and associated ground failure. US Geol Surv Open-File Rep 84-760:231–238
Youd TL, Keefer DK (1981) Earthquake-induced ground failure in facing geologic and hydrologic hazards – earth science considerations. US Geol Surv Prof Paper 1240-B:23–31
Woodroffe CD (1990) The impact of sea-level rise on mangrove shorelines. Prog Phys Geogr 14(4): 483–520

In Situ Settling Velocities and Concentrations of Suspended Sediment in Spill Plumes, Øresund, Denmark

Karen Edelvang

17.1
Introduction

The fixed link between Denmark and Sweden, the Øresund Link, consists of a combination of a double-track railroad and a four-lane highway. Starting on the Danish side, the link begins with a tunnel section, then crosses an artificial island and ends with a high bridge crossing the Sound to Sweden (The Feedback Monitoring Programme for the Fixed Link Across Øresund 1997). The link facilitates road and sea traffic between the two countries, thereby promoting the social and economic growth of the Øresund region. Strict environmental considerations accompany the construction of the Øresund complex. In no manner whatsoever are the construction activities allowed to influence and thereby modify the natural physical, chemical and biological environment of the adjacent Baltic Sea. During the building phase, only short-term environmental damage is allowed, provided it is restricted to areas close to the construction sites.

One of the aims of the environmental impact assessment is to evaluate the effectiveness of the so-called zero solution agreement concerning suspended matter fluxes through the Øresund. As a consequence, sediment spills from dredging activities have to be minimised. Monitoring and control programmes based on a co-operation agreement between Denmark and Sweden are designed for this specific purpose. All results are gathered in the Feedback Monitoring Centre of the Øresundskonsortiet's monitoring and supervision unit. Computer models are an integral part of the monitoring programme. This chapter presents some of the results obtained from field surveys during dredging activities dedicated to collecting information to validate the computed scenarios.

17.2
Study Area

The Sound between Denmark and Sweden is a strait connecting the Baltic Sea with the North Sea (see Fig. 17.1). In the shallow parts of the Sound the seabed consists mostly of lag deposits or till. Fine-grained material accumulates in the deeper areas (Kuijpers *et al.* 1991). The concentration of naturally suspended sediment is generally very low in the Øresund (<10 mg l^{-1}) showing systematic seasonal variations (Valeur *et al.* 1996). Algae dominate in the summer, and in the winter stormy episodes resuspend sediment deposited on the bed. During dredging operations large amounts of fine-grained material, mainly composed of chalky mud with no cohesive properties, are suspended in sediment plumes with concentrations which are up to 100 times higher than the natural background concentration. As the settling velocities of the individual particles are

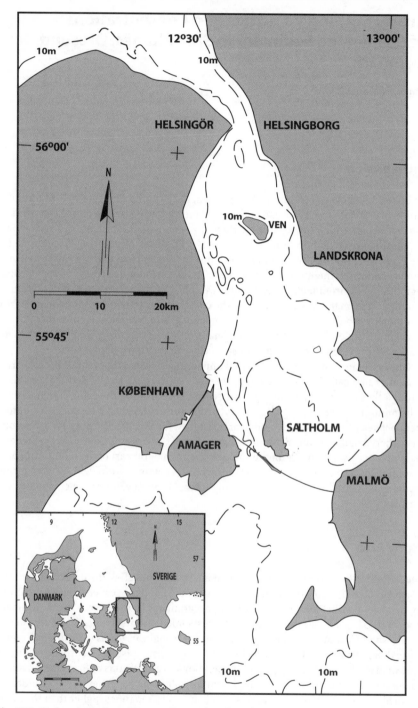

Fig. 17.1. Study area: the Sound between Denmark and Sweden

very small, such sediment clouds can be traced for several kilometres in the Sound. The sediment mainly accumulates in areas known as natural sinks for fine-grained material (Pejrup and Larsen 1994).

17.3
Methods

Median settling velocities of the spilled sediment were calculated on the basis of field investigations using a Braystoke SK 110 settling tube to measure the in situ settling velocity. The measured settling velocities were compared with settling velocities determined from sediment fluxes to the seabed (Valeur *et al.* 1996).

Suspended sediment concentrations were measured during dedicated field surveys and were also obtained from the constructor who determines concentrations on a routine basis in the course of construction. Only final quality-assured data were used. Concentrations were measured with a GMI type TU150-IR turbidity sensor calibrated by water samples collected in the field. The water samples were filtered through pre-weighed Millipore 0.45 μm CEM filters. The filters were dried at 60 °C for 2 h and the suspended sediment concentration determined.

Sediment accumulation at specific sites in the Øresund was monitored using a Haps core bottom sampler, which extracts samples down to 30 cm below the surface of the seabed. Comparisons between measured and modelled sediment concentrations and sediment accumulations were used to validate the sediment dispersal model MIKE 21 HD, PA/MT (DHI software). Grain sizes of the deposited sediment were analysed in the laboratory using a Malvern Laser Particle Sizer, which covers a particle size range from 0.2 to 600 μm.

17.4
Results

17.4.1
Settling Velocity

The suspended spill material in the Øresund has median particle sizes of about 6 μm as determined by the Malvern Laser Particle Sizer. As it consists of mainly fine chalky material it does not flocculate, and therefore has very low settling velocities. In situ settling velocities of the sediment suspended during dredging operations were computed on the basis of several settling tube analyses using Stokes' law. Measurements were carried out in sediment plumes from two different dredgers. One was a dipper dredger (Chicago) and the other a cutter suction dredger (Castor). Although the two dredgers have very different modes of operation the measured settling velocities from the sediment spill were quite similar. Thus, the median settling velocity of the sediment spilled from Chicago at a temperature of 10 °C was 1×10^{-5} m s^{-1} compared to 6×10^{-5} m s^{-1} from Castor at a temperature of 15 °C (values extrapolated to the 50% percentile). The average median in situ settling velocity was determined to be 3×10^{-5} m s^{-1} (see Fig. 17.2). Earlier investigations of vertical particle fluxes determined from sediment traps (Valeur *et al.* 1996) report a mean computed settling velocity of 4×10^{-5} m s^{-1}, which compares well with the results of this study.

Fig. 17.2. Settling velocities measured at dredging sites (Chi-cago T = 10°C; Castor T = 15°C)

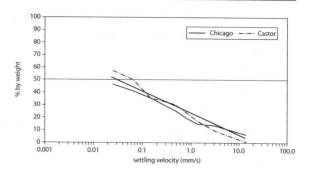

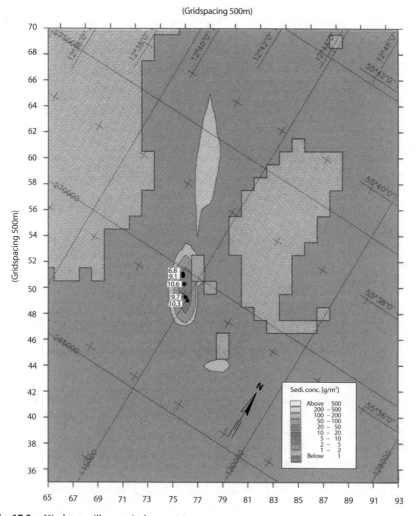

Fig. 17.3 a. Hindcast spill scenario from 14 May 1996

17.4.2
Concentration

The natural concentration of suspended sediment in the Øresund is very low, values of
1 to 2 mg l⁻¹ being the norm (Pejrup and Larsen 1994). During dredging, concentrations can rise to several hundred milligrams per litre in the area surrounding the dredger
depending on currents and the dredged material, decreasing rapidly downstream of
the source. Suspended sediment concentrations were measured on two different occasions, on 14 May 1996 and 2 October 1996. Figure 17.3a,b show plots of model scenarios
based on MIKE 21 (DHI software). The measured values are marked by dots. The data
were used to test the consistency of the model results. Concentrations are seen to vary

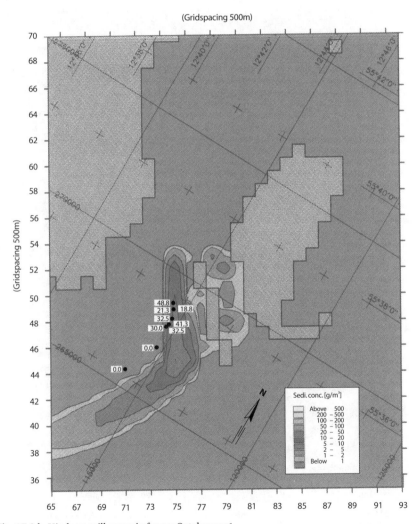

Fig. 17.3 b. Hindcast spill scenario from 2 October 1996

between 0 and 49 mg l⁻¹. In Fig. 17.3a all the sampling points are consistent with the hindcast plots. The suspended sediment concentration varies from 0 to 10 mg l⁻¹ in the hindcast compared to 6 to 11 mg l⁻¹ in the measured data points. Thus, the measured values compare well with those of the hindcast, both with respect to the concentration levels and the position of the plume. In Fig. 17.3b the simulated concentrations are again very close to the measured concentrations, ranging from 0 to 50 mg l⁻¹ in the hindcast and 0 to 49 mg l⁻¹ in the measured data. The position of the model plume is close to the actual plume. Control samples taken outside the plume area showed concentrations of <1 mg l⁻¹, supporting the evidence that the model prediction is valid.

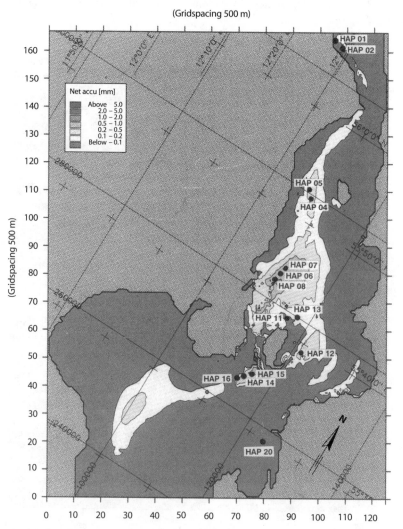

Fig. 17.4. Hindcast scenario of accumulation rates during total dredging period from 19 October 1995 until 12 December 1996

17.4.3
Accumulation

Fine-grained sediment is assumed to be deposited in the Øresund region because the deeper areas are known to act as sinks for suspended matter (Madsen and Larsen 1986). Accumulation of spilled sediment was monitored based on a hindcast scenario (see Fig. 17.4) for the whole dredging period 19 October 1995 to 12 December 1996. A total of 181 000 tons of material was spilled up to this date (Bernitt and Simonsen 1997). Samples were collected at 14 sites chosen to be representative of areas of accumulation on the basis of the scenario. The locations of the sites were chosen such that they would cover the potentially highest accumulation areas. Accumulated material, which was concluded to originate from dredging spill based on colour and texture, was found only at three of the 14 sites; the sites in question are marked as HAP 06, HAP 07 and HAP 12. These localities were found within the high accumulation areas predicted by the model. They are situated in areas sheltered from wind and wave action at water depths below 10–15 m where fine sediments are known to be deposited (Pejrup and Larsen 1994). The recovered material consisted of a 2-mm-thick layer of unconsolidated, light-coloured sediment very different from the consolidated bed. Fig. 17.5 shows a photo of the core HAP 12. Because the newly deposited material is very loose in structure, its density is close to that of seawater: 1 mm in Fig. 17.4 corresponds to 1000 kg m^{-2}. The accumulation rates given by the hindcast are 0.2–1 mm for the investigated sites for the total spill period from 19 October 1995 to 12 December 1996. This is in good agreement with the observations, even though the accuracy of this method is rather low. Generally speak-

Fig. 17.5. Bottom sample taken in an accumulation area. The light-coloured sediment at the top is partly spill

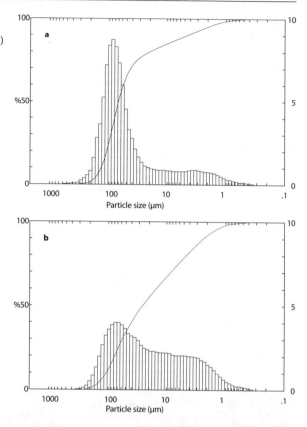

Fig. 17.6. Grain size distributions of surface samples from cores HAP 06 (**a**) and HAP 12 (**b**)

ing, the results document that, with the information used, the model actually simulates areas of accumulation rather than calculating actual accumulation rates. Possible accumulations of spill outside the predicted areas were not tested.

Surface samples from HAP 06 and HAP 12 were analysed in the laboratory. The median particle size was 70 µm for sample HAP 06 (Fig. 17.6a), being equivalent to a settling velocity of about 4×10^{-3} m s^{-1}. The median diameter for HAP 12 (Fig. 17.6b) was 25 µm, which corresponds to a fall velocity of 6×10^{-4} m s^{-1}. In both cases the median particle sizes are much larger than those measured for the spill material from the dredgers. The positive skewness for both analyses indicates a population of fine material of about 6 µm, which may very well be derived from spill. Mineralogical analyses have shown that only about 10% of the material deposited at HAP 06 was derived from spill plumes (Knudsen and Lomholt 1997). The much larger particle sizes thus indicate that coarser natural sediments were also deposited.

17.5
Discussion and Conclusions

The settling velocities measured in the areas surrounding the dredgers are the same for the different types of dredgers. The average median settling velocity is computed to

3×10^{-5} m s^{-1}, which is relatively slow. This is because chalk deposits from the Danian (Tertiary), which consist of small bryozoan fragments, dominate the bed material at the dredging sites. This material does not flocculate. Settling velocities will diminish downstream of the source, because the particles with the largest settling velocities will settle first. The individual fine particles can be transported over long distances before they settle.

Concentrations of the suspended material close to the dredgers are generally very high, but these are rapidly diluted downstream of the source. Measurable concentrations can be traced for more than 10 km. The concentration of suspended matter does not influence the measured settling velocities as no flocculation takes place.

Investigations of possible accumulation areas show that sediment spill from dredging can be traced to specific sites where fine-grained material is normally deposited. Such areas occur at water depths below 10–15 m. Bed sediment containing spill material generally consisted of a light-coloured fluffy layer, about 10% of which originated from dredging operations.

Acknowledgements. The data used in this chapter originate from the Feedback Centre; Øresundskonsortiet has kindly given permission to use it. The author would like to thank Jens R. Valeur, Anders Jensen and Poul Hammer for discussions and assistance.

References

Knudsen C, Lomholt S (1997) The Øresund Link: Recognition of fine-grained sediments from excavation activities – pilot study –. Danmarks og Grønlands Geologiske Undersøgelse, Rapport 1997/114

Kuijpers A, Larsen B, Nielsen PE (1991) Surface sediments in the Danish part of the Sound. DGU Map Series no. 26, Copenhagen

Bernitt L, Simonsen J (1997) Hindcast no. 9. Feedback Monitoring DHI/7783/90 0269, Øresundskonsortiet, Technical report

Madsen PP, Larsen B (1986) Accumulation of mud sediments and trace metals in the Kattegat and the Belt Sea. Report of Marine Pollution Laboratory no. 10. Charlottenlund, Denmark

Pejrup M, Larsen B (1994) Natural sediment transport through Øresund. DHI/7170, Øresundskonsortiet, Technical report

The Feedback Monitoring Programme for the Fixed Link Across Øresund (1997) DHI/7783-8, Øresundskonsortiet, Technical report

Valeur JR, Pejrup M, Jensen A (1996) Particle dynamics in the Sound between Denmark and Sweden. ASCE Conference Proceedings, Coastal Dynamics '95: International Conference on Coastal Research in Terms of Large-Scale Experiments, pp 951–962

Index

traffic management 37
trailer suction hopper dredger 160
transformation capacity 136
Troll gas field/terminal 83, 157

U
UK Department of the Environment 38
UK Environment Act 50
UK Marine Foresight Panel 56
UK Water Resource Act 50
UK Wildlife and Countryside Act 51
UN Universal Declaration of Human Rights 29
UN World Charter for Nature 29
urbanisation 141
US Council on Environmental Quality 92
US Environmental Protection Agency (EPA) 92
US National Environmental Protection Plan 92
US National Oceanic and Atmospheric Administration (NOAA) 92

V
valuation approach 146
Vinci, Leonardo da 9, 10, 13
Voluntary Marine Reserve 59

W
Wadden Sea 17, 21, 25, 28
Wadden Sea National Park 17, 26, 27, 29, 83, 118, 119, 158
waste disposal 56, 153, 155
water quality 37, 42, 45, 55, 59, 68, 70, 111, 137, 152
water scarcity 91
water storage 37
weir 35
wetland 25, 28, 117, 119, 146, 147, 149, 178
World Heritage Convention 28
World Heritage Site 60
WWF 26

Z
Zeepipe pipeline 157
zero dumping concept 122
zero solution 105, 106, 181

Springer
and the
environment

At Springer we firmly believe that an international science publisher has a special obligation to the environment, and our corporate policies consistently reflect this conviction.
We also expect our business partners – paper mills, printers, packaging manufacturers, etc. – to commit themselves to using materials and production processes that do not harm the environment. The paper in this book is made from low- or no-chlorine pulp and is acid free, in conformance with international standards for paper permanency.

Springer

Printing: Mercedesdruck, Berlin
Binding: Buchbinderei Lüderitz & Bauer, Berlin